INTRODUCTION TO PRACTICAL MARINE ENGINEERING

Volume1: Text

by

Alan L. Rowen
Professor Emeritus of Marine Engineering, Webb Institute

Raymond F. Gardner
Professor of Marine Engineering, US Merchant Marine Academy

Jose Femenia
Professor of Marine Engineering, US Merchant Marine Academy

David S. Chapman
Associate Research Scientist, University of Delaware

Edwin G. Wiggins
Professor of Marine Engineering, Webb Institute

ISBN 0-939773-48-1

PREFACE AND ACKNOWLEDGMENTS

This book was conceived to meet an evident need for an introductory textbook for new marine engineering students who had little, or more likely, no prior experience. It was inspired by Reno King's WWII-vintage book Practical Marine Engineering that served so many of us well, and like that book, most of the chapters began as the authors' lecture notes. When it was decided to share the material more widely additional chapters were added. Throughout, we tried to keep the language simple and concise, with an emphasis on nomenclature and very basic principles, and with the assumption that the readers may not have had even a high-school course in physics.

The two volumes are divided intentionally to enable a reader to examine a figure while simultaneously reading the text. Tables for Chapter 1 and Chapter 21 are incorporated with the figures for the same reason, while the remaining tables are embedded in the text.

Many of the illustrations are the original work of the authors and appear here for the first time. Others were provided by manufacturers who very generously permitted their reproduction for this book, or previously, for use in other SNAME publications. Other illustrations were extracted from publications of the United States Naval Institute and from other sources as noted. Portions of this material have appeared in different form in Modern Marine Engineer's Manual, 2nd and 3rd Editions, published by Cornell Maritime Press, and in SNAME publications.

The authors acknowledge with gratitude the efforts of Dr. Richard C. Harris, who edited much of the text, and Anton DePasquale, Timothy Yen, Jonah Rowen, and Susan Evans, all of whom assisted in preparing the figures for publication.

Alan Rowen
New York City
December, 2004

CONTENTS

CHAPTER 1
INTRODUCTION TO THE MARINE POWER PLANT

1.1 Purpose

In general, the power plant of a ship or other waterborne vessel provides both for its propulsion through the water, and for **ship's services**. Ship's services include propulsion engine support (for example, fuel and lubricating-oil service), steering gear, navigation and communication equipment, the **hotel load** (lighting, heat, ventilation, running water, provision refrigeration), and support of cargo, trade, or mission requirements (such as cargo heating, cranes for cargo handling, elevators and powered ramps on ferries, and weapons and sensors on a warship). The plant is therefore likely to include not only the **main propulsion** engine, but also electric generators, pumps, compressors, heat exchangers, tanks, piping, control systems, and equipment for maintenance and repair.

1.2 Historical context

The first steam ships, propelled by paddle wheels driven by single-cylinder reciprocating engines, with the steam generated in wood- or coal-fired kettle boilers, were operating on rivers and lakes in the early years of the nineteenth century. By the late nineteenth century, triple-expansion reciprocating steam engines driving propellers enabled long-distance voyages to be undertaken, more reliably and in less time than even the fastest sailing ships. Boilers had become safe and reliable, although coal was still the principal fuel, requiring large crews of hard-working stokers on larger ships. By the turn of the century, both the steam turbine and the diesel engine had been introduced, and electric generators, powered steering gear and deck machinery, water-distillation equipment, and mechanical refrigeration were commonplace. By the beginning of WWI, steam turbines were preferred over reciprocating engines for higher-powered ships, and oil had replaced coal as the principal fuel.

In the years after WWI, concerted efforts, mostly in the United States, to develop steam turbines and the necessary mechanical gearing, and boilers capable of supplying high-pressure, superheated steam, paved the way for both merchant and naval ship power plants through WWII and into the 1970s.

Diesel engines used fuel more efficiently than steam plants, but were more labor intensive. Diesel engines therefore found greater favor in Europe than in the United States. Nevertheless, diesel engines of modest power levels developed rapidly after WWI, inspired in part by the need for a reliable engine for submarines and by a large market for stationary power and locomotives ashore. After WWII, advances in diesel engine performance enabled them to compete with steam turbines for increasing shares of the merchant ship sector.

The reciprocating steam engine was obsolete before WWII, supplanted by steam turbines for high-powered installations even before WWI, and then by diesel engines at lower power levels. It was only as a matter of expedience that reciprocating steam engines were revived during the early part of WWII, as the emergency construction of ships outpaced diesel engine, turbine, and gear-building capacity, leaving reciprocating steam engines to be fitted in the Liberty ships, and in many corvettes and escort aircraft carriers.

The decades after WWII saw the introduction and evolution of the gas turbine and of nuclear power, and the continued evolution of steam turbine plants and diesel engines. After the rapid escalation in fuel costs of the middle 1970s, diesel engines became the dominant form of propulsion for merchant ships, naval support ships, and smaller warships, while gas turbines came to dominate the medium-sized and larger warship sectors.

1.3 Current status

Marine power plants in common use today include **steam turbines**, **diesel engines**, **gas turbines**, and **nuclear plants**. The reciprocating steam engine can be considered obsolete. Gasoline engines are widely used in small boats and pleasure craft but are rare in other applications.

1.4 A typical oceangoing merchant ship

Two examples are offered to illustrate the relation of the power plant to the ship that it serves. The first is a modern, oceangoing merchant ship, shown in Fig. 1.1. The power plant is located in an **engine room** or **machinery space** at the aft end of the hull, thereby locating the propulsion machinery close to the **propeller**, while leaving most of the hull free for carrying cargo. The deck house or **accommodation block**, which provides living and office space, is surmounted by the **wheelhouse** or **bridge**, and is located above the engine room, leaving most of the **weather deck** clear for cargo or cargohandling. The **engine casing**, which may be incorporated into the accommodation block or may be a separate structure as shown here, continues the engine room up to the base of the **smokestack** or **funnel**. The exhaust ducts of engines and boilers are carried through the casing and stack, enabling the exhaust gases to be released high enough to be carried clear of the ship. The narrow space aft of the engine room, low within the hull, is the **after peak**, usually used as a seawater **ballast tank**, and surrounds the **stern tube**, which carries the **propeller shaft** from the engine room to the propeller. The space above the after peak, continuing aft to the **transom**, is reserved for the **steering gear**, which must be directly over the **rudder**, and for storage, often including freshwater tanks. The **double bottom** extends through the engine room, and forms a solid foundation for the major machinery.

Typical merchant ship speeds are in the range of 12 to 20 knots, requiring propulsion power ranging from less than a few thousand horsepower in the smaller ships, to over 60,000 horsepower in the largest, fastest ships. Except where shallow draft limits the size of the propeller, this requirement can be met with a single propeller, as in this example. Such single-screw ships have the most efficient hull forms and are cheaper to construct and operate. Likewise, concentrating the machinery in a single engine room simplifies construction, operation, and maintenance.

The crew size is small, fifteen-to-twenty-five people, with perhaps four licensed engineers and two or three certified mechanics. The machinery is fully automated and normally operates under remote control from the bridge, with the engine room unattended except during emergencies or during critical periods, for example, when entering or leaving port. The engineers work nearly normal hours, spending their workdays in routine tasks, including maintenance and paperwork. At the end of the workday, one engineer is designated to make night rounds of the machinery, and to be the first to respond to alarm conditions. This contrasts to crewing practices of the recent past, when automation was much more limited, and two- or three-man watches were stood around the clock.

In general, a merchant ship must make a profit for its owner, and it is therefore worked hard: port stays are minimized, and the ship spends most of its time under way at sea, at close to full power. It is not unusual for merchant ships to be at sea for over 300 days a year. The ship is rarely taken out of service for maintenance, at most for one week out of the year, as this represents a financial loss to the owner and a disruption of schedule. Overhaul opportunities are limited, and reliability of the machinery is important, although reliability must be weighed against cost.

This typical merchant ship power plant may be characterized as economical in initial cost and operation, and therefore relatively simple to build, operate, and maintain, reasonably reliable, and fuel efficient. Weight and volume of the machinery are usually of secondary importance.

1.5 A typical medium-sized warship

Fig. 1.2 is a plan view of the hull of a modern warship of destroyer size, with a beam of about 16 meters. To achieve a maximum speed of about thirty knots, power requirements are on the order of 80,000 to 100,000 horsepower, which must be divided between two propellers. To improve the survivability of the ship in combat, the machinery spaces are divided.

Because of the high ratio of power to ship size, machinery weight and volume are of primary importance, as is reliability. Additional requirements might be imposed, including resistance to shock to improve survivability, and limits on noise and vibration. The machinery tends to be complex and maintenance intensive when compared to merchant ship machinery. The ship is intended to alternate periods of service at sea with periods in port, when extensive maintenance can be performed by specialists based ashore.

When a warship is at sea, full power is required for only a small part of the time; most of the time is spent at a low power level, using perhaps a tenth of the power installed, yielding a modest **cruising** or **endurance speed** of about fifteen knots, and reducing fuel consumption and the strain on the machinery. The propulsion machinery must be designed with the ability to operate satisfactorily at both full power and cruising-power levels. In some ships, separate engines of modest power are installed to meet the vastly different cruising-power requirement.

The warship power plant can thus be characterized as light and compact for the power achieved, high in reliability and survivability, but complex, initially expensive, and costly to maintain and operate. Fuel efficiency is most important in terms of range, since only limited quantities can be carried.

1.6 Fuels

The overwhelming majority of marine power plants use liquid petroleum **fuel oils. Distillate fuel oils (DO)** are the easiest to use but are the most expensive, and **residual fuel oils** are the cheapest but the most difficult to use. **Intermediate fuel oils (IFO)** are produced by blending distillate fuels with residuals, and generally have all of the difficulties of residuals although to a reduced extent. The density of the distillate fuel is the least, so that it is often referred to as **light fuel**, with blended and residual fuels all being **heavy fuel oils (HFO)** . The heaviest fuels must be heated to reduce their **viscosity** before they can be pumped, and all heavy fuels must be heated further before they can be burned. Heavy fuels generally contain impurities, including solid particles, water, sulfur, vanadium, and sodium, and tend to leave deposits of carbon during combustion, all of which contribute to the deterioration of the machinery. Distillate fuels contain little, if any, of these impurities, do not generally require heating, and do not normally leave carbon deposits. Even in power plants in which heavy fuels can be used, the choice of fuel is mostly an economic decision, requiring that a balance be struck between the lower cost of the heavier fuels and the inconvenience and increased costs of fuel treatment and machinery maintenance. The power plants of most sea-going merchant ships are run on heavy fuels, but most coastal, river, and harbor craft, fishing boats, and naval vessels of all types, use distillate fuels.

Coal is used in some trades, where its much lower cost, well below that of even residual fuel oils, justifies the difficulties inherent in its use. Coal-fired ships are necessarily steam ships, with coal storage arrangements, coal transfer systems, and boilers all specifically designed for the coal. Nevertheless, there are indications that coal will be used as a marine fuel more frequently in the future.

Natural gas is the principal fuel for liquefied natural gas carriers, and its use in coastal, river, and harbor craft is increasing. Where natural gas is available, it is an attractive fuel that burns cleanly in boilers, diesel engines, and gas turbines, reducing maintenance costs and exhaust emissions of atmospheric pollutants. Difficulties associated with natural gas for marine use are in the storage and handling arrangements.

Because of the reduced exhaust emissions, natural gas use in coastal, river, and harbor craft is likely to increase.

Nuclear fuel is the dominant choice for the largest surface warships and for long-range submarines. In general, nuclear power is no longer considered viable for merchant ships, for economic and social reasons, the only exception being a fleet of Russian Arctic icebreakers.

1.7 Combustion

Petroleum, coal, and natural gas are **hydrocarbons**, comprised principally of hydrogen and carbon. In combustion the carbon and hydrogen rapidly oxidize, combining with the oxygen in air, and releasing considerable heat. The heat released is the energy that is then converted in an engine to provide mechanical power. This heat, per unit mass of fuel, is the **heating value** of the fuel, about 18,000 Btu/lb or 42,000 kJ/kg for fuel oil, and about 12,000 Btu/lb or 28,000 kJ/kg for coal.

Air is primarily a mixture of oxygen and nitrogen. To completely oxidize the constituents of the fuel, reducing the carbon to carbon dioxide and the hydrogen to water vapor, requires about fifteen pounds or kilograms of air for every pound or kilogram of fuel oil, the theoretical or **stoichiometric air-fuel ratio**. For coal, the stoichiometric air-fuel ratio is about ten. If insufficient oxygen is available, or if the fuel is not adequately mixed with the air, combustion will be incomplete, the full heating value of the fuel will not be released; **carbon deposits** will remain behind, and smoke will be generated. To ensure complete combustion of oil fuels, they are **atomized**, or broken into small droplets, prior to combustion, and **excess air** is supplied, above the stoichiometric quantity. In oil-fired marine boilers 10 to 15 percent excess air is typical, while in diesel engines about 100% is used. In marine practice complete combustion of coal is achieved by burning the coal on a grate in the presence of a steady flow of air from beneath.

The products of combustion, ultimately exhausted to atmosphere, include, along with the carbon dioxide and water vapor from the oxidation of the carbon and hydrogen in the fuel, the nitrogen drawn in with the air, and the oxygen remaining from the excess air. In addition, other elements in the fuel are represented in the exhaust, including sulfur, sodium, and vanadium oxides, all of which can contribute to deterioration of equipment with which they come in contact. The sulfur oxides, in particular, along with trace quantities of nitrogen that oxidize at high temperatures, are considered atmospheric pollutants.

1.8 Comparison of steam, diesel, gas turbine, and nuclear plants

Table 1.1 compares some of the features of propulsion plants in current use. The following notes are meant to clarify the contents of the table:

The power range is given per unit for the steam turbines. One unit normally drives one propeller shaft. As many as four units might be used in a large, fast ship, with four propellers, developing about 300,000 shp, as on the big aircraft carriers.

Low-speed diesel engines are usually arranged with each engine driving one propeller shaft. A twin-screw ship would therefore have two engines.

Medium- and high-speed diesels, and **aircraft-derivative gas turbines**, are frequently fitted in pairs, geared together, with each pair driving a single propeller shaft, as in the warship described above.

Heavy-duty gas turbines could be arranged in pairs but, in fact, all of the ships that were built with these engines had single-engine plants.

LNGCs are liquefied natural gas carriers; **naval auxiliaries** are non-combatant naval vessels, including tankers, stores ships, tenders, and many amphibious-force ships.

The line for fuel quality shows the poorest quality fuel that is normally used in each type of plant. Especially for the diesel engines, there is an inverse relation between fuel quality and maintenance, so that, if a better quality fuel were used in a medium-speed engine, for example, maintenance requirements would be lower.

Major maintenance of aircraft-derivative gas turbines is usually done by replacing the unit in service with a rebuilt unit after several thousand hours of service. This practice reduces maintenance at sea but adds to the shore-side support requirements.

For the nuclear plants, much of the machinery is inaccessible for maintenance when the plant is in use.

1.9 Prominent features of steam and diesel plants for a merchant ship

Prominent physical features of steam and diesel plants, as they might be fitted in the single-screw merchant ship of Fig. 1.1, are shown in the capsule schematic drawings of Fig. 1.3, all drawn to the same scale. This discussion is intended to provide a broad overview of typical propulsion plants: equipment introduced here is described in detail in later chapters. Gas turbine and nuclear plants are not included, since they are less likely to be found in this application, but analogous information is included in Chapters 18 and 19.

These views are partial **inboard profiles**, viewing the engine room from a point just to starboard of the ship's centerline, looking to port, so that only machinery on the centerline, or to port, is visible. Some features are common to all three plants:

There is a deep double bottom below the engine room, providing a solid foundation for the propulsion machinery.

The propeller shafting exits through the after engine room bulkhead. At the forward end of the propeller shafting is the **main thrust bearing**. The thrust bearing transfers the thrust of the propeller to the hull of the ship, and is therefore always in line with the propeller, and solidly mounted.

Each boiler, or each engine, has an **uptake** carrying exhaust gas through the casing and up the smokestack.

In the steam ship, one of the two boilers is visible (the other is on the starboard side), with its **forced-draft fan** to supply combustion air. Steam passes from the **boilers** to the **turbogenerators**, and also to the **high-pressure turbine** (to starboard of centerline), and then to the **low-pressure turbine**, from which it exhausts downward to the **condenser**. High- and low-pressure turbines drive the propeller shaft through the **reduction gear**. The gearing, together with the thrust bearing, is mounted on a solid foundation, which is in fact, integral with the double bottom. The **deaerating-feed tank** is part of the system by which water is returned from the condenser to the boilers. Since these prominent components of the plant are connected by piping, there is considerable flexibility in the way that they can be arranged: the vertically stacked arrangement in Fig. 1.3, which minimizes engine room length, is common in merchant ships.

For ships with low-speed diesel engines, the main engine itself often defines the size and shape of the engine room, as Fig. 1.3 shows. In this single-screw ship it is on the centerline, directly connected to the propeller shaft, and directly mounted on the double bottom: there is almost no practical alternative to this arrangement. A thrust bearing is built into the aft end of the engine. Exhaust from the main engine rises through a **waste-heat boiler**, to provide steam for ship's services; electricity is supplied from **diesel generators**. The overall volume of the engine room is larger than that of the medium-speed diesel plant, or of the steam plant.

Medium-speed diesel engines are more compact than low-speed engines. In the example of Fig. 1.3, two engines, side-by-side, drive the propeller shaft via reduction gearing. The engines and gearing, together with the thrust bearing, are mounted together on a foundation that is integral with the double bottom. As with the low-speed diesel plant, a waste-heat boiler and diesel generators provide for ship's services.

1.10 Energy distribution and losses; shaft power and brake power

Only a small percentage of the energy in the fuel is ultimately converted to useful power. Most of the energy in the fuel is dissipated to the surroundings as heat:

1 to 2% is lost to surroundings from hot surfaces

some heat is rejected with the exhaust:

about 10% in steam plants
25 to 30% in diesel plants
65 to 70% in gas turbine plants

some heat is rejected to the sea, via cooling water:

in steam plants, mostly at the condenser, up to 60%
in diesel plants, at engine coolers, about 30%
in gas turbine plants, less than 1%, at lubricating-oil coolers

The balance is **brake power** output of the plant or engine, and, in steam plants, energy supplied to ship's services:

about 30% in steam plants
up to 45% in diesel plants
about 30% in gas turbine plants

This value is the **thermal efficiency** of the plant, and is inversely related to the fuel consumption: as marine propulsion plants, diesel engines are the most efficient, and have the lowest fuel consumption.

Except for the low-speed diesel, the brake power is provided at too high a speed for driving the propeller, and is therefore reduced in speed, usually by gearing. About 1 to 2% of the brake power is dissipated in the gearing. What is passed on is the **shaft power**.

There are further losses in bearings, perhaps another 1%, but there is a large loss at the propeller itself, which, under the best circumstances, might convert only about 60 to 70% of the shaft power to thrust. As a result, the **thrust power** will be 30% or less of the fuel energy in a diesel plant, and 20% or less in a steam plant or gas turbine plant.

1.11 Power requirements vs. ship speed

The **cube law** is a rule of thumb that indicates that the power required to propel a ship is proportional to the ship speed cubed. It is more accurate for ship speeds of about twenty knots or less, but nevertheless, it provides a useful indication of the high cost, in terms of power and therefore of fuel consumption, of high ship speeds. For example, if a merchant ship were to serve a trade at 19 knots instead of 15 knots, it would require a power plant about twice as powerful, and would consume twice as much fuel. As another example, the warship capable of a maximum speed of 30 knots at full power would require only an eighth of its power to cruise at 15 knots.

CHAPTER 2
STEAM PLANT OVERVIEW

2.1 The steam cycle

A simple **steam cycle** is illustrated in Fig. 2.1. In the **boiler**, fuel, usually oil, is burned, and the heat released is used to generate **steam**. The steam, which is at high **pressure** and **temperature**, and which therefore has tremendous energy, is directed to blow against the blades of a **turbine**, spinning its output shaft with a force sufficient to drive the propeller of a ship or to drive a generator. This steam exhausts from the turbine at low pressure and temperature that reflect the fact that most of the energy that the steam had when it left the boiler was converted to power in the turbine. The exhaust steam is collected in a **condenser**, where enough of the energy remaining in the steam is removed to cause it to condense. This water, called **condensate**, is then pumped back to the boiler, as **feedwater**. The fact that the same water is used over and over again, as feedwater, steam, and condensate, is what justifies the term **cycle**.

Some of the principles and terminology just introduced will be explained in more detail in the text that follows, but some questions, likely to occur to someone encountering the steam cycle for the first time, are worth answering immediately:

> One question likely to arise is about how a steam turbine works. The simple answer is that it works like a windmill: rows of windmill-like **blades** are attached to a shaft, which spins when the steam is directed at the blades through **nozzles**.
>
> Another likely question concerns the apparent waste involved in cooling the steam to condensate at the condenser, only to then reheat it, first as feedwater and then further in the boiler. In fact, condensation is necessary for the cycle to work, partly because the power required to pump the feedwater up to the high pressure needed to force it back into the boiler is much less than what would be required to compress the low-pressure steam to this high pressure; so much power would be required to compress the steam, that it would exceed the power developed in the turbine, rendering the cycle unworkable.
>
> A third question concerns the condenser. The cooling water used in the condenser is drawn from, and returned, to the sea. It never mixes with the steam that it condenses, so that the condensate remains pure, suitable for use as feedwater.

2.2 A typical steam ship engine room

Figs. 2.2a, b, and c show the arrangement of the machinery in the engine room of a typical steam ship. Usually, one enters the engine room at the main deck level and proceeds down the ladder to the **fireroom** on the **operating flat** forward (i.e., to the right, in these drawings) of the boilers. The capsule descriptions which follow should be taken one-by-one, with equipment named located in any of the three drawings in which it appears. Further information on the items named or visible in the drawings can be found through the index.

> The fireroom is at the level of the furnaces of the boilers. This ship is typical in having two boilers, and in burning fuel oil, designated as FO in these drawings.
>
> The fuel oil is stored in **bunker tanks** built into the ship's hull in several locations. It is transferred daily from these bunker tanks to the **FO settling tanks** in the engine room by the **FO transfer pump**. In this example, the FO settling tanks and transfer pump are visible on the lower, or **floor-**

plate level, of the engine room. From the settling tanks, the fuel is delivered to the boiler furnaces by the FO **service pumps**, via strainers and heaters.

The air needed for combustion of the fuel is supplied to the furnaces by the **forced-draft fans**, which are located above the boilers.

Most of the steam produced in the boilers is passed to the propulsion turbines via the **throttle valves** operated from the main console. The propulsion turbines require so many rows of the windmill-like blades mentioned above, that they are divided into the two, separately encased units, the **high-pressure (HP) turbine** and the **low-pressure (LP) turbine**, through which the steam passes in series.

The turbine shafts spin too fast to drive an efficient propeller. The usual way to reduce the turbine shaft speed is with a **reduction gear**, which also serves to combine the power produced by the HP and LP turbines.

The power output of the reduction gear is carried aft, through the **thrust bearing**, and finally through the hull of the ship, to the propeller, by the **propeller shaft**. The forward part of the propeller shaft is visible in the drawings, supported by one of several **steady bearings**. The propeller forces the ship through the water by developing a thrust, or force, which is transmitted along the length of the shaft to the thrust bearing, which transfers the thrust to the hull of the ship.

Most of the steam that is not used in the propulsion turbines flows to the **turbogenerators**. Each turbogenerator consists of a steam turbine, a reduction gear, and an electric generator. The electricity produced is distributed through **switchboards** to meet all of the ship's electricity requirements, including the motors, lighting, and control equipment in the engine room.

Exhaust steam from the propulsion turbines flows to the **main condenser**, below the LP turbine, and exhaust steam from the turbogenerators flows to the **auxiliary condensers**. (These adjectives, **main** and **auxiliary**, are applied fairly consistently: main is usually associated with propulsion, as in main reduction gear, and main condenser; all other equipment is auxiliary, as in auxiliary condenser. Many auxiliaries are fitted in duplicate.)

In the condensers, the steam is condensed on the outsides of tubes, which are kept cool by seawater, supplied by the **circulating pumps**. These pumps draw the water from the **sea chests** visible at floor-plate level, which are open to the sea. The condensation of the steam is made to occur under a vacuum, which must be maintained by the **air ejectors**, visible on the operating level. The condensate that results must be drawn out of the condensers, against this vacuum, by the **condensate pumps**. (Note the use, in the drawings, of the main and auxiliary designations.)

Miscellaneous drains, mostly from heaters, collect in the **atmospheric drain tank**, from which it is drawn by the **drain transfer pumps** to join the flow from the condensate pumps. This combined flow passes to the **first-stage feed heater** and then to the **deaerating-feedwater heater**, both of which may be seen in the elevation. From the deaerating-feed heater, the flow, now called feedwater, flows to the **feed pumps** on the lowest level, which will pump it to the boilers. The feed pumps are often the only auxiliaries other than the generators to be driven by steam turbines, rather than electric motors.

All bearings require lubrication and cooling. Some bearings, like the steady bearings for the propeller shaft, and those of most auxiliaries, are self contained, but for the propulsion turbine bearings, the main reduction gear, and the thrust bearing, a main lubricating-oil system is fitted: **lubricating oil**

(LO) collects in a sump below the reduction gear, from which it is drawn by the **LO service pumps**, and pumped, via strainers, through the **LO cooler**, to a **LO gravity tank** (visible in the elevation), from which it flows to the bearings and gear teeth. The **LO purifier**, **LO settling tank**, and the **LO storage tank** are all associated with this system.

Compressed air is used for many purposes aboard ship, including numerous control systems, supplied from the **control-air compressor** via the **control-air receiver** and dryer. Air for other purposes is supplied from the **ship's-service air compressor** and receiver.

The **distilling plant**, alternatively called an **evaporator**, a **freshwater generator**, or a **desalination plant**, is used to produce freshwater from the sea. The distilling plant is complex, with connections for steam, seawater, and freshwater, and accounts for a number of the small pumps and heat exchangers in its general vicinity. Some of the freshwater produced is stored in the **distilled-water tank**, for use as **make-up feed**, to compensate for losses from the steam cycle. The rest of the freshwater is stored in the **potable-water tanks**, for washing, cooking, and drinking. The potable water is distributed for use throughout the ship by the **freshwater pumps** and **pressure tank**, some of the water passing first to the **hot-water heater**.

Refrigeration and **air-conditioning machinery**, comprising compressors, receivers, condensers, and brine systems, with associated pumps, tanks, and heat exchangers, is fitted for provision refrigeration (the SS refrigeration unit in the figure), air conditioning of crew's quarters, and, in this example, cooling of some cargo spaces.

The space at the bottom of the engine room, below the floor plates, and similar spaces elsewhere, as in the cargo holds and shaft alley, are called **bilges**. **Bilge pumps** are fitted to empty water which accumulates in these bilges. Where the water may be contaminated by oil, it must first be passed through an **oily-water separator** (not shown) before being discharged overboard. **Ballast pumps** are fitted to move water into and out of designated hull tanks, in order to maintain the stability of the ship and to control its draft and trim. The vacuum priming equipment is used with the bilge and ballast pumps.

Fire pumps are fitted to supply seawater to hydrants throughout the ship, one of several fire-fighting systems provided.

No attempt is made here to identify all of the equipment shown in the figures; again, reference should be made to other sources, including later chapters of this book.

2.3 Properties of steam

Consider an uncovered pot of freshwater placed over a flame. As the water absorbs heat from the flame it will come to a boil at 212°F, and steam will be generated. As long as the pressure remains at normal atmospheric pressure (generally taken as 14.7 psi, or pounds force per square inch) neither the water nor the steam will rise above 212°F. This temperature is the **saturation temperature** for water at this pressure and is a unique property for the water. If the flow of steam is restricted by a cover on the pot, the pressure will rise, and the covered pot can be properly called a boiler. As the pressure rises, so will the saturation temperature: at 100 psi, the saturation temperature is 328°F. In many marine steam plants steam is generated at about 1000 psi, at which the saturation temperature is 545°F.

When water is boiling, the steam and water are both at the same pressure and saturation temperature, but they represent different **phases** of water, and differ in properties, including density and energy. The energy difference between the saturated steam and the saturated water is the **latent heat**. The water boils because some molecules have absorbed the latent heat of evaporation and therefore change phase.

In using steam for producing power, the pressure is important, as it is the difference in pressure of the steam, at the inlet to a turbine, relative to the pressure at the exhaust, which enables the steam to do work. The amount of work that the steam can do depends on its energy, and the energy, in turn, depends on the temperature as well as the pressure.

Steam at the saturation temperature is called **saturated steam**. If it is trapped and conducted away from the saturated water, it can be further heated, or **superheated** with respect to its saturation temperature. Fig. 2.3 is a schematic representation of the superheating process. In the marine plants cited above, where steam is generated at 1000 psi, with a saturation temperature of 545°F, the steam is commonly superheated to 955°F. The principal reason for superheating steam is to raise its energy content, commonly measured in **British thermal units**, or **Btu**. There are several forms of energy used by engineers, but for our immediate purposes, the applicable form is the **specific enthalpy**, or the energy content of one pound mass of steam, in Btu/lbm. Saturated steam at 1000 psi and 545°F has an enthalpy of 1192 Btu/lbm, but when the steam is superheated to 955°F, its enthalpy rises to 1480 Btu/lbm. Relative to saturated steam at the same pressure, therefore, the superheated steam has a much greater energy content.

It is true that a great quantity of low-pressure, low-energy steam could be used to develop the same amount of power as a smaller quantity of superheated, high-pressure steam, so that a comparison should also be made of the volume of steam at different conditions, since the size of the machinery can be compared on the basis of the volume of steam required. Engineers refer to the **specific volume** of the steam, which is the reciprocal of its density, and is measured in cubic feet per pound mass. For saturated steam at 100 psi the specific volume is 4.4 cuft/lbm, while the specific volume of superheated steam at 1000 psi and 955°F is 0.8 cuft/lbm. Taking into account the greater energy of each pound of high-pressure superheated steam, as well as its much lower volume, it follows that the equipment used to generate the required amount of steam, and then to convert the energy to power, can be much smaller, and therefore, lighter and less expensive. While it is not so obvious, it is also true that the high-pressure, superheated steam plant will operate much more efficiently.

To conclude this brief introduction, therefore, it can be said that, while low-pressure, saturated steam is useful for heating, and may be adequate for low levels of power production, when levels of power sufficient for ship propulsion are involved it is desirable that the steam be generated at an elevated pressure and superheated to a high temperature.

2.4 Feedwater

Think back to the pot of boiling water. For the process of steam production to be continuous, as water evaporates, it must be replaced at the same rate. In a boiler generating steam under pressure, then this **feedwater** must be pumped in at a higher pressure. This is the job of the **feed pump**.

Another fact to consider is that whenever water is boiled the vapor is essentially pure and all of the impurities remain behind. Eventually a pot or boiler would become encrusted with scale. Furthermore, at high saturation temperatures, a boiler would be subject to corrosion because of oxygen present in the feedwater, and also because the water turns slightly acidic at these temperatures. To prevent these problems the feedwater is highly distilled (initially, in the ship's own distilling plant), deaerated (partly in the deaerating-feed heater), and chemically treated to suppress remaining oxygen, to maintain slight alkalinity, and to inhibit scale formation. Perhaps most important is the fact that almost all of the steam generated is recovered after use, and condensed, to be reused as the boiler feedwater, with only enough make-up feed introduced to compensate for losses.

CHAPTER 3
BOILERS: BASIC FEATURES

3.1 Introduction

It is the function of a boiler to produce steam. The normal heat source is fuel, burned in a furnace in combination with air. The earliest boilers were little more than kettles placed above brick furnaces. As pressures rose above what kettle-type containers could withstand, and as it became obvious that steam could be produced more rapidly if the container were small in relation to the furnace, marine-propulsion boilers evolved in the direction of the modern **watertube boilers**, of which the most typical is the two-drum type. A simple example of a **two-drum boiler** is the small, saturated steam boiler illustrated in Figs. 3.1 and 3.2, while Figs. 3.3, 3.4, and 3.5 show large propulsion boilers.

The principal components of the small boiler of Figs. 3.1 and 3.2, most evident in Fig. 3.2, are as follows:

> **generating tubes**, filled with water, which constitute the heat-absorbing surface, heated from the furnace or by the hot gases leaving the furnace, from which water absorbs sufficient heat to change phase;
>
> drums, the st**eam drum** at the top and the **water drum** at the bottom, which are large-diameter manifolds, to which the ends of the generating tubes connect;
>
> the **furnace**, in which the fuel is combined with air and burns; in this boiler the furnace is lined entirely with **refractory** (firebrick, in the figure) except for the side wall, which is covered by **water-wall tubes**;
>
> the **casing**, usually of sheet steel with layers of insulation, which encloses the boiler to make it gas tight, provides connections for air supply and exhaust, and limits heat loss to the surroundings;
>
> supports, which transfer the weight of the boiler to supporting structure.

The larger boilers of Figs. 3.3 and 3.4 have the following additional components:

> **superheater tubes**, into which the saturated steam flows to be superheated;
>
> **headers**, which are manifolds smaller in diameter than drums, to which are connected those tubes which are not connected directly to the drums;
>
> **screen tubes**, which are generating tubes placed between the superheater and the furnace.

The function, design, and construction of these elements will be discussed, but it is first necessary to understand some principles of feedwater and circulation.

3.2 Feedwater and circulation

Recall the pot of boiling water used as an example in Section 2.3. Eventually all of the water will evaporate and the bottom of the pot, no longer cooled by water at the saturation temperature, will overheat. As the temperature of the metal rises, its strength diminishes, until it fails. This problem can be prevented by maintaining a flow of water to the pot, matched to the flow of steam that leaves. The water in the pot, never rising above the saturation temperature, will absorb heat from the metal as rapidly as the metal absorbs it from the flame, preventing the metal temperature from rising to dangerous levels.

This same principle applies to every section of steam-generating surface in a boiler. Feedwater is supplied to the steam drum by a feed pump, at a rate that keeps the water level in the drum at about the half-way point. This water must then circulate through the generating tubes, in a process illustrated in Fig. 3.6: the tubes closest to the furnace, exposed to the heat radiated directly from the flame as well as the highest temperature of the gases of combustion, are heated most intensively, and are therefore generating steam at a high rate. On the other hand, the tubes farthest from the furnace are heated only by the passing gas, much cooler at this point, and are generating little steam. Because the density of the steam is much lower than that of the water, the average density of the mixture of steam and water in the tubes nearest the furnace is lower than the overall density of the mixture in the tubes farthest from the furnace. Consequently, the heavier mixture in the farthest tubes will flow down to the water drum, forcing the lighter mixture in the closer tubes to flow upwards into the steam drum. This pattern of flow is called **natural circulation**, because it occurs only as a result of the difference in density, and therefore, in weight, of the mixture in the tubes. It should be emphasized that all of the water and steam in all of the generating tubes is at essentially the same temperature, i.e., the saturation temperature, and that the difference in density occurs because of the difference in proportion of steam and water.

The tubes in which the mixture of steam and water flows upwards are called risers, and those in which the flow is down, are **downcomers**. It is possible to so intensively fire a boiler that even the farthest tubes are generating significant quantities of steam, and the difference in density is insufficient for adequate circulation. To preclude the damage that would then result, additional, **external downcomers** are used, located outside of the gas path, and therefore never heated. External downcomers are normally fitted between the steam and water drums of propulsion boilers, and are visible in Fig. 3.4, where they are completely external to the casing. (External downcomers are also used to supply water to groups of tubes that are not connected at the bottom to the water drum, but are instead supplied via headers, an arrangement also visible in Fig. 3.4.)

Generally speaking, all marine propulsion boilers are naturally circulated. (When a pump is used to create the flow, the process is called **forced circulation**. Forced circulation is common for marine waste-heat boilers, but otherwise, only for very high-pressure boilers ashore.)

3.3 Generating tubes

Generating tubes are all of those in which steam might be generated. They can be divided into groups:

Most generating tubes are in the generating bank, connected to the water drum at the bottom and the steam drum at the top. Even in the small boiler of Fig. 3.2, these constitute only part of the generating surface.

Water-wall tubes are present in Fig. 3.2 as side-wall tubes, and also appear in Figs. 3.3, 3.4, and 3.5, where they are run from headers at the bottom, supplied by either direct connection to the water drum, or by downcomers from the steam drum. In Fig. 3.3, it can be seen that the side-wall tubes bend at the top to become roof tubes, and then enter the steam drum directly, while the rear-wall tubes rise to a header, which is then connected to the steam drum by the L-shaped risers visible above the roof of the furnace.

Screen tubes are found in boilers with superheaters, where they protect the superheater from the heat radiated directly from the flame. In Fig. 3.3, three rows of screen tubes can be seen, running from the water drum to the steam drum. In Fig. 3.4, again there are two rows running to the steam drum, but at the bottom, these tubes turn at a right angle, away from the water drum, and run under the floor of the furnace to the header visible at the bottom.

Tubes in the generating bank and screen are in rows, perpendicular to the gas flow. In the direction of gas flow, they may be **in-line** or **staggered**, as in the generating bank of the boiler of Fig. 3.2. While staggering promotes slightly better heat transfer, in-line tubes are easier to keep clean on the gas side (called the fireside) and tend to be preferred today.

Water-walls may be formed of **tangent tubes** or as **welded wall**, both illustrated in Fig. 3.7. Welded wall forms a largely gas-tight structure, sufficiently strong to be self-supporting, needing only **buckstays** (see Fig. 3.5) to prevent bowing, and requiring only minimal casing. Welded wall is therefore the preferred approach today.

All generating tubes that are adequately circulated will be well cooled, with tube-metal temperatures not much above the saturation temperature. They are therefore made of carbon steel, generally by an extrusion process that is referred to as **seamless cold-drawing**. Generally the tubes are secured in the drums or headers by **cold rolling**, also called **expanding**, in which the hole is reamed to a diameter only slightly greater than that of the tube, so that when the tube is inserted in the hole, a tool comprised of hardened steel rollers can be inserted and used to expand the diameter of the tube, within the thickness of the drum or header, sufficiently to form a strong, leak-proof interference fit.

3.4 Drums and headers

Drums and headers are manifolds, tying together the ends of the many tubes, to form continuous circuits and to provide structural cohesion. In addition, the steam drum has the following functions:

to collect the steam-and-water mixture leaving the risers and to provide space for the separation of the water from the steam before the steam leaves the drum;

to act as a reservoir (or **surge tank**) for the water contained in the boiler (a surge tank permits the water level to rise and fall without adversely affecting operation);

to receive the feedwater, allow it to mix with the steam-and-water mixture leaving the risers and so bring the incoming feed quickly to saturation temperature, and distribute it to the downcomers.

The water drum and lower headers have the following additional functions:

to distribute water received from the downcomers to the risers;

to act as receptacles for accumulations of solids that precipitate from the circulating water.

The upper headers serve to collect the steam-and-water mixture leaving those risers not directly connected to the steam drum, so that it may be passed to the steam drum.

Drums differ from headers principally in size, the drums being large enough for a person to enter, via a **manhole** at one end, while a header will usually have a row of **handholes** opposite the tube entrance. The construction of manholes and handholes is illustrated in Figs. 3.8 and 3.9.

To retain strength under pressure, despite the many tube holes, drums, and, to a lesser extent, headers, must be very thick-walled. Headers are usually single-piece forgings, but drums, especially steam drums, are generally fabricated of steel plate, rolled to shape and welded, as illustrated in Fig. 3.10. This figure also shows that the **wrapper sheet**, without tube holes, can be much thinner than the **tube sheet**. The ends of the drums, called **heads**, are separate steel forgings, welded on as in Fig. 3.8. Because the drums and generating-circuit headers are outside the gas path, they will not rise above the saturation temperature, and can be made of carbon steel.

3.5 Superheaters

Superheaters consist of U-shaped tubes, run between headers, as in Fig. 3.3, where the tubes are horizontal and the vertical headers are visible at the rear of the boiler, and in Figs. 3.4 and 3.5, with vertical tubes connecting at the bottom to horizontal headers. The ends of the **U-tubes** opposite the headers must be supported, usually by clips welded to adjacent generating tubes, which then become **superheater-support tubes**. The clips are visible in Fig. 3.3. To permit access for cleaning and repair, the space inside the U of the innermost tubes is about 18 inches wide, forming a **walk-in cavity**, accessible, when the boiler is shut down, through a bolted door in the casing.

Superheater tubes are cooled only by the steam, which is not only hotter than the saturation temperature, but does not remove heat from the tube walls as well as boiling water would. Great care must therefore be taken in design to prevent overheating of the superheater tubes. Usually:

Expensive high-temperature alloy steels are used.

Screen tubes are placed in front of the superheater to protect it from the heat radiated directly from the flame.

Internal bulkheads in the headers, called **diaphragms**, force all of the steam to flow successively through groups of tubes in series, as in Fig. 3.11, so that the flow rate is higher than if the steam were allowed to flow through all of the tubes simultaneously. Each group of tubes is called a **pass**, and the superheater of Fig. 3.11 is therefore a seven-pass unit. The greater flow rate of the steam improves its ability to absorb heat from the tubes.

Saturated steam is brought from the steam drum to the superheater inlet connection on one of the headers by an external pipe. Depending on the number of passes, odd or even, the superheater outlet connection may be on the same header.

3.6 Furnaces

The furnace of a boiler must be large, in order to:

mount a sufficient number of **burners**;

contain sufficient air for complete combustion, typically over 200 cubic feet for each pound of fuel burned;

accommodate the whole flame of each burner, without allowing the flame to impinge on the furnace lining or screen tubes.

Furnaces that are completely lined on the walls and roof with refractory are not used in modern propulsion boilers; instead, water-wall tubes are used to line the furnace, for the following reasons:

Furnace refractory requires considerable maintenance, largely because of the expansion-contraction cycles that result from normal operation, as a boiler is brought into service, then later shut down, and also because of the vibration and shock of marine service. Water-wall tubes require little attention.

The water-wall tubes add highly productive generating surface, with little added weight or volume. In many boilers the water-walls generate as much steam as all of the other generating tubes.

Refractory furnace floors, however, are common, for these reasons:

Refractory stays in place on the floor and requires little maintenance.

The refractory protects the floor tubes from the corrosive effects of **slag**, which accumulates during combustion as a molten mass formed from incombustible constituents of the fuel.

The floor tubes are horizontal or have only shallow slopes, and so do not circulate well. In addition, any steam generated rises to the upper halves of the tubes, which is the surface exposed to the heat of the furnace. Where floor tubes are present, therefore, they are usually protected by refractory.

Refractory-lined furnace surfaces must be supported from a structure, usually of steel plate, supported in turn from the drums and headers. The bricks are the inner lining, backed by layers of insulation, all attached to the furnace plate by inserted anchor bolts. Tangent-tube and welded-wall surfaces can be self-supporting, with layers of insulation mounted externally, secured by sheet metal screwed to clips welded to the tubes.

The burners may be located in the front of the furnace, as in Figs. 3.1 and 3.3, or the top, as in Fig. 3.4, the choice depending most on space constraints imposed by the installation, which may determine furnace proportions.

At junctions between the floor and the walls there is a sloping **corbel**, made in-place of **plastic refractory** (unfired refractory cement, which hardens when the furnace is fired). In Fig. 3.3 the corbel at the base of the rear wall is visible. The corbel helps to make the furnace gas tight, protects the headers from the slag, and also protects them from the heat radiated from the flame. Plastic refractory may also be used to form **gas baffles**, by being packed over an armature of wire laced around tubes. Gas baffles are visible in Fig. 3.3 at the top and bottom of the screen tubes, where they guide the gas away from the drums and also protect them from the heat of the flame. (Gas baffles can also be formed by welding steel bar stock into the space between tubes, creating the same effect as the welded wall of Fig. 3.7.)

3.7 Casings and supports

The casing encloses the boiler and includes the **windbox** at the furnace, where the burners are mounted and the air supply is connected, and the gas exit, where connection is made to the uptake. Most of the casing consists of sheet metal panels, which may carry additional insulation, bolted to a framework, which is supported in turn from the drums and headers, from furnace plates, or from water-wall tubes.

In single-casing construction, most common with welded-wall furnaces, the sheet metal securing the insulation can serve as the casing, as in Figs. 3.2 and 3.7. A double casing, clearly shown in Fig. 3.3, is formed when the steel panels are supported at a distance from the furnace enclosure to form an outer casing, with the outside of the furnace enclosure forming the inner casing. The void space of a double casing may be used as part of the air duct from the forced-draft fans to the windbox, thereby limiting gas leakage and heat loss to the surroundings.

The windbox is a steel fabrication that in effect, forms an enlarged double casing in way of the burners, at the front or top of the furnace, as in Figs. 3.3 and 3.4, respectively. The windbox allows the air from the forced-draft fans to be distributed to the burners, which are usually mounted at the inside to the furnace wall or roof, and at the outside, to the windbox enclosure.

Boilers of the type discussed here are most conveniently supported at the bottom, as in Fig. 3.4, from a **saddle** welded to each end of the water drum and, below the furnace, to each end of a lower side-wall or floor header. The saddle is the top of the **footing**, which is bolted to a **pedestal** built up from ship's structure. Where no such header is present, the furnace floor structure will provide mounting for the furnace footings. Usually, only one of the footings is fixed in place in both horizontal directions, the others being bolted through slots or enlarged holes, so that they may slide sufficiently to permit thermal expansion. This is the reason for the slope of the furnace footings in Fig. 3.4.

CHAPTER 4
BOILERS: HEAT RECOVERY EQUIPMENT

4.1 Introduction

The efficiency of a boiler is largely determined by the amount of heat remaining in the exhaust gas, and each decrease of 10°F in exhaust gas temperature represents about a quarter percent rise in boiler efficiency. However, as the gas leaves the last tubes of the generating bank, it will still be some 100°F above the saturation temperature, too hot to exhaust economically, but not hot enough for further heat transfer to the boiling water.

There are two means used today to enable this heat to be recovered:

- Preheat the feedwater on its way to the steam drum, in an **economizer**.
- Preheat the combustion air supplied by the forced-draft fan, on its way to the windbox, in an **air heater**.

4.2 Cold-end corrosion

Unfortunately, there is a lower limit to the temperature to which the exhaust gas can be safely cooled, because of the normal presence in the gas of sulfuric acid vapor, derived from the small quantities of sulfur present in almost all oil fuels. As long as the acid remains in its vapor phase it is harmless, but in the presence of surfaces at about 300°F or less the vapor will start to condense, and corrosion of the surface will begin. Because this problem occurs at the exhaust temperature, which is low compared to combustion temperatures, the term **cold-end corrosion** is applied.

4.3 Economizers

The boilers in Figs. 3.3 and 3.4 are both fitted with economizers. A section of an economizer is shown in Fig. 4.1. The configuration of all three of these examples is similar and typical: an upper, water-inlet header, and a single pass of a dozen or more tubes, running back and forth across the uptake passage, through which the water flows down to a lower outlet header. The 180-degree turns between the straight segments of the tubes are usually forged separately and welded in place. Because of the relatively low temperature difference between the gas and water, the gas-side surface area of the tubes is generally augmented by circular fins. The flow of water is downward, counter to the upward flow of gas, so that the hottest water is heated by the hottest gas.

With this configuration, it follows that at the top of the economizer the coolest gas is heating the coldest water, so that the top rows of tubes would be subject to cold-end corrosion if the feedwater were too cool. Because the tubes are subject to high internal pressure any reduction in metal thickness caused by corrosion would soon result in local failure of the tube, with unacceptable rates of water leakage, which would force a shutdown for repair. Consequently, feedwater systems are designed to ensure that the temperature of the water entering an economizer is not lower than about 280°F. The resulting gas exit temperature will be about 315 to 320°F, and the boiler efficiency, about 88.5 to 89 percent.

4.4 Air heaters

The only type of air heater in common use for marine propulsion boilers today, for recovery of exhaust gas heat, is the **rotary-regenerative air heater**. Figs. 4.2 and 4.3 show a typical example, and Fig. 4.4 shows how the unit relates to a boiler and forced-draft fan.

As shown in the figures, the unit consists of a flat casing spanning both the air duct and the uptake, and enclosing a vertical cylindrical drum, which is slowly rotated about a vertical axis by an electric motor. The drum is a steel framework mounting pie-shaped baskets of closely-spaced, partly-corrugated plates, which form many small passages for the air or gas. As the drum rotates, these plates are heated by the gas in way of the uptake, and then, as the drum rotates into the air duct, the heat is removed by the relatively cool air.

The air supplied to a boiler is normally drawn from the upper reaches of the engine room by the forced-draft fan, so that its temperature will be about 100°F as it enters the air heater. Conditions are set for cold-end corrosion, but while corrosion of the plates will gradually reduce performance, it will not normally result in a failure requiring a shut down. The design of the unit is such that components subject to cold-end corrosion are renewable. Because corrosion is tolerable if not avoidable, these units are usually designed to reduce the temperature of the exhaust gas to values well below those acceptable with economizers, as low as 260°F, resulting in boiler efficiencies of 90 percent or more.

4.5 Summary

The choice of heat recovery device is a trade-off between the smaller bulk and lower maintenance requirements of the economizer, and the higher efficiency of boilers fitted with rotary air heaters. It should be noted that the heat recovery device affects other parts of the system: when economizers are fitted, steam-air heaters are usually fitted to preheat the combustion air, which improves combustion. When rotary air heaters are used, additional feedwater heaters are designed into the cycle.

It is not uncommon to find large boilers ashore fitted with both an economizer and an air heater, but in marine plants, given the generally lower power levels, the constraints on space and weight, and the need to operate over a range of power levels, the added initial cost and greater complexity would be difficult to justify. For marine boilers, therefore, it is usually a choice of one or the other.

CHAPTER 5
BOILERS: ACCESSORIES

The accessories described here are fittings, mountings, and other components normally associated with boilers, which are not commonly encountered elsewhere. This is by no means an all-inclusive list.

5.1 Internal fittings and mountings

Boiler internals are fitted inside the steam-and-water space of the boiler, mostly in the steam drum, as in Fig. 5.1. Those described here are the most important.

Internal-feed pipe: a horizontal, perforated pipe that distributes the incoming feedwater along the length of the steam drum. The feed pipe is mounted about a quarter of the drum diameter above the bottom of the steam drum, and penetrates the head of the drum to connect to the feedwater piping from the feed pump.

Surface-blow pipe: a horizontal, perforated pipe located just below the normal water level in the steam drum, which penetrates the head of the drum to connect, via the surface-blow valve, to the bottom-blow piping which leads to the bilge or overboard. When the surface blow valve is opened, boiler pressure will blow out the upper layer of water in the steam drum, and with it any impurities, including oil, which had floated to the surface.

Steam baffles and **cyclone separators**, which have the function of separating the water and steam in the mixture entering the steam drum from the generating tubes, so that the steam can leave with minimum moisture, while the water combines with incoming feed to enter the downcomers. When baffles are used alone, as shown in Figs. 3.3 and 5.1, they are usually installed just below the normal water level, as a nearly horizontal, triple layer of perforated plates. The perforations are staggered, forcing the rising mixture into a tortuous path, so that water droplets are flung off. A cyclone separator is shown in Fig. 5.2, and two are visible in the steam drum of Fig. 3.4. Cyclone separators are used in high-performance boilers: solid baffles are fitted in the drum to direct the water and steam mixture from the generating tubes to the shell of the cyclone, where it enters tangentially. The tangential entry gives the mixture a swirl, forcing the heavier water to the periphery of the shell, where it can drain down over vanes at the bottom, while the steam rises through a central hole at the top to exit through the scrubber.

Dry pipe: a horizontal pipe located at the top of the steam drum, running the length of the drum, perforated along its upper surface, and connected by a welded tee to the saturated steam outlet, which rises through the wrapper sheet. Steam exiting the drum must make a final sharp turn to enter the dry pipe through the perforations, flinging off any remaining water droplets.

Desuperheaters: horizontal loops of piping fitted in the steam drum below the water level, or in the water drum, with connections at both ends passing through the heads of the drum, as in Fig. 5.3. Superheated steam is passed through the desuperheater, from one connection to the other, and is thereby cooled by exchanging heat with the surrounding saturated water in the drum. Most boilers equipped with superheaters have two desuperheaters, one, an **auxiliary desuperheater** supplied with steam from the superheater outlet, which delivers desuperheated steam via a valve at its outlet, for driving auxiliary services, and the other, a **control desuperheater**, used to control the temperature of the superheated steam. The auxiliary desuperheater is usually located in the steam drum, but the control desuperheater may be in either drum.

Chemical-feed pipe: a small, horizontal, perforated pipe that runs the length of the steam drum, below water level, from a connection that penetrates the head of the drum. The pipe is used to inject water-treatment chemicals into the boiler. The same pipe is used at other times to extract a sample of the boiler water for chemical testing.

Swash plates: partial bulkheads in the steam drum, which limit sloshing of the water caused by ship motions. The plates provide convenient support for some of the other internals.

5.2 External fittings and mountings

Boiler external fittings and mountings are located outside the steam-and-water space, and are, in fact, generally visible on the outside of the casing. The items described in this section are the most important, but do not include instrumentation and automation, or burners, which are discussed in later sections.

Safety valves: spring-loaded valves set to open at a pre-determined pressure, discharging steam to atmosphere, in order to prevent the boiler pressure from exceeding design limits. A typical safety valve is shown in Fig. 5.4. Two such safety valves are mounted to connections on the steam drum wrapper sheet, and a third at the superheater outlet. The safety-valve outlet connects to piping led up the smokestack. The superheater outlet safety valve is set to open first, thereby maintaining a flow of steam through the superheater even if the drum safety valves open subsequently. (If the drum safety valves opened first, they could divert steam away from the superheater, allowing the superheater tubes to be overheated.) With reference to the figure, the compression screw is used to adjust the popping pressure at which the valve opens, after which the cap is secured in place by the boiler inspector to prevent tampering. (In this example, which follows Navy practice, a padlock is used; in merchant ships, the cap is secured with twisted wire and a lead seal).

Gage glasses: fitted to show the water level in the steam drum. A typical example of the type suitable for boiler pressures is shown in Fig. 5.5. They are mounted, usually in adjacent pairs, to connections on the steam drum, above and below the normal water level at mid-height of the drum. Because the water chamber is connected to the drum at both ends, the water level will be reproduced. A lamp behind the glass makes the level visible.

Stop valves: used to isolate the boiler from connected piping when the boiler is not operating on the line. There are **feedwater-stop valves** in each feed line (most boilers are fitted with two parallel feed lines as a safety precaution), located before the economizer inlet header when an economizer is fitted; a **main-steam stop valve** at the superheater outlet, but after the branches to the auxiliary desuperheater and the connections for the superheater safety valve and the circulating vent; and an **auxiliary-steam stop valve** at the desuperheater outlet.

Air vent: a valve at the top of the steam drum, also called a drum vent, open directly to atmosphere, used to release air from the steam drum when the boiler is lit off, and then closed; the air vent will also be opened to let air in when the boiler is shut down and almost cold, or when the boiler is cold and being drained, or to release air when the boiler is cold and being filled (cold boilers are usually kept full to minimize internal corrosion).

Superheater-circulating vent: a valve at the superheater outlet, with its outlet piping led up the smokestack, opened before the boiler is lit off, and left open until the boiler is operating on the line, so that, as steam is produced, it flows through the superheater, keeping the tubes from overheating. The circulating vent is opened when preparing to shut the boiler down and is left open until the boiler is cold.

Bottom-blow valve: connected at the bottom of the water drum, with its discharge led to the bilge and overboard; used to force water and precipitated contaminants out of the boiler. For generating circuits that do not drain to the water drum, additional bottom-blow valves are fitted at the lowest header.

Header drains and vents: fitted respectively to lower and upper headers, including those of the superheater and economizer, as required to ensure complete draining or venting.

Feedwater check valves, **feedwater regulating valves** and **feedwater regulators**: used to maintain the water level at mid-height in the steam drum. Feedwater check valves are manually operated, and are fitted in the feed line upstream of the automatic feedwater regulating valve, for which they are an emergency replacement. The feedwater regulating valve is operated by the feedwater regulator, which senses the level in the drum, and may also sense the rate of steam flow, and adjusts the valve accordingly.

Sootblowers: used to blow loose deposits from the firesides of the tubes. An example of a manually operated, rotary-element, steam sootblower is shown as Fig. 5.6. The head, on the left, is outside of the casing, while the element is inside, between or adjacent to rows of tubes, and extends across the gas passage to be supported in a bearing on the outside of the casing opposite. When the chain wheel is turned by pulling the chain, the element rotates, as does the cam. When the cam opens the valve, steam, usually supplied from the desuperheater outlet, flows through the gooseneck to the nozzles, and as the element rotates, sweeps the tubes. When the sootblower is not in use, air from the forced-draft fan discharge flows through to keep the element cool. Sootblowers of this type are used in the generating bank and economizer, but in the superheater, the gas is too hot for air cooling to suffice, and retractable-element types are used. A pneumatically operated example is prominent in Fig. 3.3: the pneumatic operator drives the element into the superheater cavity and rotates it simultaneously, so that the steam, emitted from nozzles at the end, sweeps the tubes in a helical path. In use, the element is cooled by the steam; it retracts from the gas path when not in use.

5.3 Burners

A typical oil-fired burner is illustrated in Fig. 5.7. It is comprised of the centrally mounted oil atomizer assembly and surrounding **air register**. The unit mounts through the windbox, with the innermost component in the figure, the register throat ring, secured to the furnace wall. The outer part, carrying the **air doors**, **impeller plate**, and **atomizer**, is bolted to the outside of the windbox. The air flowing into the windbox enters the furnace through the **throat**, some of it flowing through the slots of the impeller plate, giving the air a swirling motion. The atomizer introduces the fuel into the center of this swirling air stream as a fine mist of oil droplets, which also have a swirling motion, but opposite to the swirl of the air. This opposition results in good mixing of the oil droplets with the air, a prerequisite to good combustion. Once this mixture is ignited it will burn continuously as long as the air and oil flows are maintained.

The business end of a mechanical atomizer is shown in Fig. 5.8. Oil is supplied through the hollow barrel under pressure, which forces it through the oil passages to the tangential slots, and then through the orifice, from which it emerges as a swirling mist of fine droplets. In steam atomizers, this process is assisted by steam.

While auxiliary boilers may have only a single burner, propulsion boilers have at least two, to facilitate control, and to permit a fire to be maintained even while changing atomizers, which can be withdrawn even when the boiler is in service. To vary the firing rate, the pressure of the fuel to all of the burners is adjusted by a regulating valve in the piping external to the burners. The amount of air is correspondingly adjusted, usually by a **damper** at the forced-draft fan inlet. Individual burners can be taken out of service by shutting off the fuel supply and closing the air doors, or, on a rise in demand, they can be returned to service (**burner sequencing**). When the boiler is cold, the outer part of the burner may be withdrawn, as in Fig. 3.4, or, if mounted on hinges, swung aside, to provide access to the furnace.

5.4 Instrumentation and automation

The following instruments are commonly fitted to propulsion boilers:

pressure gages for:
- steam drum
- superheater outlet steam
- fuel, before and after the pressure regulator
- feedwater, before and after the economizer

draft gages or **manometers** for:
- forced-draft fan discharge
- windbox
- furnace
- uptake, before and after the economizer or air heater

thermometers, or other temperature indicators, for:
- fuel
- air, before and after the air heater
- gas, before and after the air heater or economizer
- feedwater, before and after the economizer
- superheater outlet steam

water gage glasses
periscope or other smoke indicator
flame detector

The following alarms are commonly fitted to propulsion boilers:

- high or low steam pressure
- high superheater outlet steam temperature
- high or low water level
- high uptake temperature
- low feedwater pressure
- low fuel pressure
- low fuel temperature
- low forced-draft pressure
- ignition failure
- smoke
- flame failure
- high feedwater salinity (an indication of impure feedwater)
- fire in windbox or uptake
- loss of electric power for control circuits

The following automatic controls are commonly fitted to propulsion boilers:

safety valves
automatic combustion control, to maintain the superheater outlet pressure as steam demand changes, by adjusting the firing rate and by burner sequencing, and to maintain the proper air-fuel ratio for good combustion

automatic feedwater regulator, to maintain an adequate water level in the steam drum, by regulating feedwater flow

automatic superheat temperature control, to maintain constant superheater outlet steam temperature, usually by diverting some of the steam through the control desuperheater, then returning it to be mixed with the balance

automatic fuel cut-off, which shuts off the fuel supply in the event of a flame failure, forced draft failure, or extreme low water level

CHAPTER 6
BOILER OPERATIONS

The procedures described here are for an oil-fired propulsion boiler, but are otherwise general, intended to provide background information only, and are not comprehensive or all-inclusive. Recommended procedures for specific equipment should be sought out and followed.

6.1 Preparation and lighting off

If a boiler has been shut down for an extended period or has been worked on, it must be thoroughly inspected: tools, rags, and debris must be cleared from the drums and headers, the furnace, the windbox, and the vicinity of the forced-draft fan intake. Air ducts and uptakes must be clear, with any lay-up covers removed. Manhole and handhole covers should be closed, with dogs hand tight.

Root valves for pressure gages, gage glasses, and other instruments or sensors are checked, to ensure that they are open about a half turn. The manual lifting cables for the safety valves are inspected. Steam stop valves are eased off their seats, then closed hand tight. The air vent, the superheater-circulating vent, and the superheater-header drains are all opened wide.

The water level is established at about a quarter of the lower gage glass, by draining or filling. The feed system is tested by using it to feed the boiler, even if this means draining the boiler to the bottom of the gage glass first, so that it may be filled. When main and auxiliary feed systems are fitted, both feed systems are used in turn. If there is any question about the quality of the water in the boiler, it is tested and treated. If necessary, the boiler is drained, flushed, and refilled with make-up feedwater. Bottom and surface blow valves, and all drains other than those at the superheater headers, should be tightly closed.

Fuel is lined up to the burner root valves, which should be closed. The flame monitor must be bypassed and the automatic fuel cut-off valve opened for fuel to flow. For a boiler normally operated on heavy fuel, if heated fuel is available, the recirculating valve is opened, and fuel is circulated until hot fuel reaches the root valves, when the recirculating valve is closed. If heated fuel is not available, distillate fuel must be used to raise sufficient steam to heat the heavy fuel.

Immediately before lighting off, the furnace must be purged by opening all the registers and running the forced-draft fan for at least one minute, with the damper wide open. The damper is then set for minimum air flow, and all registers except that of the center-most burner are closed. An atomizer fitted with a small tip is placed in this burner and lit with the igniter or a torch. If the fuel does not ignite immediately, or if the flame fails at any time, the fuel valve must be shut and the furnace purged before lighting off again. As soon as a stable flame is achieved, the flame monitor and automatic cut-off are activated.

6.2 Raising steam and cutting in on the line

After the center-most burner has been burning for about ten minutes, another atomizer, with a small tip, is inserted in an adjacent burner, and the flame transferred, again using a torch or igniter, as the new atomizer may not ignite directly from the first. This rotation of the burners should continue for up to one hour.

After the boiler is fired for about thirty minutes, steam will begin to issue from the air vent, which may be closed after another ten minutes. Steam pressure will then rise more rapidly. During this period, as increasing amounts of steam are generated, the water level will swell towards the middle of the steam drum without adding feedwater. The firing rate may not require adjustment, but it is usually possible to reduce the amount of excess air as the furnace warms.

Atomizers are prepared with tips sized to meet the load expected when the boiler is cut in. The boiler is cut in on the auxiliary steam line first, then on the main steam line. As the pressure approaches line pressure, cold steam lines downstream of the steam stop valves are warmed, opening drains and using bypasses around the stop valves. If bypasses are not fitted, the stop valve itself should be cracked open. When the lines are warmed, the drains are closed, the stop valves are opened wide, and the bypasses are closed. After the last stop valve is opened, the new burners are lit, and, as the boiler pressure equalizes with line pressure, it begins supplying steam to the line. The feedwater stop valves are opened and the water regulated normally.

Care should be taken that the steam pressure does not rise to lift the safety valves before the boiler is cut in. It may be helpful to shut off the burner as the pressure approaches line pressure, lighting off again only after the steam stops are opened.

When it is certain that the boiler is supplying steam to the line, the superheater-circulating vent and superheater-header drains are closed. The atomizer used to raise steam is replaced. Alarms and cut-offs are tested. Burner control is then turned over to the automatic combustion control.

6.3 Routine operation

Routine operation of a boiler requires that steam demand be met, by adjustment of the firing rate, while the water level, the steam pressure, the superheater outlet temperature, and the correct air-fuel ratio are all maintained. To achieve these goals, feed pressure, forced draft, fuel pressure and temperature must all be adequate, as well as such other variables as atomizing steam pressure. On most modern steam ships these functions are all under automatic control, but controls must be checked frequently and adjusted as necessary.

Combustion conditions may require adjustment, especially to accommodate changes in the properties of the fuel. Smoke indicates too little air, cold fuel, or a fault in the equipment, but a clear exhaust from the stack may be the result of too much air. When the fuel is at the right temperature and pressure, when the atomizers are clean, with tips properly sized and in good condition, and are properly positioned in the registers, and when other variables such as atomizing-steam pressure are correct, then the appearance of the flame may be used as an indication of air-fuel ratio:

> a bright-yellow flame, short, with sparks at the edges, with a clear furnace and a clear stack, indicates too much air;
>
> a golden-orange flame, with long trails occasionally streaked with brown, with the opposite wall of the furnace only intermittently visible, with a very slight brown haze at the stack, indicates the lowest practical air-fuel ratio.

Boiler-water chemistry must be maintained. A sample of the boiler water is tested daily. The chemistry is adjusted by injecting measured quantities of treatment compounds and by blowing down small quantities of water using the surface blow line, replacing the water removed with make-up feed.

The boiler-fireside must be cleaned by operating the sootblowers, often twice daily. Permission to blow tubes must be obtained from the bridge. While blowing tubes, excess air is increased to help carry soot up the stack.

At intervals determined by experience, cleaned atomizers are exchanged with those in service, and fuel suction and discharge strainers are switched and cleaned.

6.4 Emergency procedures

Most boiler emergencies result in a disruption of steam supply, which is especially critical during maneuvering periods. Every effort must be made to avoid disruption, to restore the boiler to service without damage, and to keep the bridge informed. In extreme cases, the safety of the ship justifies the risk of damage to the machinery.

Low water level: A low, but still visible water level should be restored to normal, but if the water level is low and no longer visible, the boiler should be secured, without further addition of feedwater, and inspected for damage before attempting to restore it to service.

High water level: If the water level is high but no longer visible in one boiler, the firing rate of the boiler should be reduced, effectively taking the boiler off the line. The surface blow line can then be used to reduce the level, and after the cause of the high water level is determined and corrected, the boiler can then be restored to service.

Water leak: A minor leak is the likely problem if a rise in use of make-up feed is accompanied by difficulty in maintaining water chemistry in one boiler. The suspect boiler should be kept in service until the consumption of make-up feed becomes unacceptable, or until an opportunity becomes available to secure the boiler. A major leak, such as a ruptured tube, is usually obvious, as it will be difficult to maintain the water level, and the leak will probably be audible, and perhaps visible through view ports. If the water level remains visible, the boiler should be secured normally, with the water level maintained. After repairs are made, or leaking tubes are plugged, the boiler can be returned to service.

Panting: Panting is a symptom of insufficient combustion air. The boiler should immediately be supplied with more air, overriding the automatic controls if necessary, while the cause of the problem is found and corrected. The problem may be caused by accumulated oil on the furnace floor, in which case sufficient air should be supplied to burn it off as cleanly as possible, while locating and correcting the cause. Care should be taken to avoid lifting the safety valves.

Forced-draft failure: On a complete loss of forced draft, the fuel supply to the boiler is cut off automatically. Air supply should be restored by locating the cause of the problem and correcting it, by opening a tripped damper, clearing an obstruction, or starting a tripped fan. If necessary, air supply can be obtained through emergency crossover ducts from the fan normally supplying another boiler, if that fan is run at a higher setting. When air supply is restored, the fuel root valves at the burners should be closed, the fuel cut-off valve reset, and the boiler purged, lit off, and returned to service.

Water in the fuel: Water in the fuel is usually evident by a drop in boiler pressure, or by sputtering at the burners. If sufficient water is present, it will extinguish the flame. The fuel pump should immediately be switched to high suction from the same settler. If the flames were not extinguished, they will continue to burn irregularly until water remaining in the line passes through the atomizers. If the flames are extinguished by the water, the fuel supply will be cut off automatically: the recirculating valve should then be opened to clear the line, then the fuel root valves at the burners should be closed, the fuel cut-off valve reset, and the boiler purged, lit off, and returned to service. Water in the settler should be drained, and the fuel pump returned to low suction.

6.5 Securing

To secure a boiler, the fires are shut off, the air registers are closed, the steam stops are closed, and immediately, the superheater circulating vent is opened. The atomizers are removed to prevent stagnant oil from carbonizing as the boiler cools. Feedwater flow must be maintained, to maintain the water level as it shrinks. Before the steam pressure has fallen to atmospheric pressure, which might take six to eight hours, the air vent must be opened. The boiler will be cool enough to enter after about twelve hours.

If the boiler is to be idle for even a few days, it should be laid up wet: the superheater-circulating vent is closed, and the boiler is filled with feedwater until water issues from the air vent, which is then closed. If the lay-up period is to be longer than a few days, a register should be withdrawn and a heater placed in the furnace, or a fan-heater should be placed in the register opening. Periodically, the waterside should be pressed up, and the water chemistry checked.

If the waterside of the boiler is to be opened, the water must be drained, with the air vent opened, to a point below the lowest manhole or handhole to be opened. When opening either the fireside or the waterside, an upper opening should be opened first, then a lower one. These precautions are particularly important within the first days after the boiler has been secured, when the contents of the boiler can still be hot enough to cause injury.

CHAPTER 7
STEAM TURBINES: BASIC PRINCIPLES

7.1 Introduction

Fig. 7.1 illustrates a simple steam turbine of a type that might be used to drive a pump. Steam is supplied to the steam chest at high pressure and temperature, hence at high energy or, as identified previously, high **enthalpy**. The steam flows through the **nozzle**, where its high enthalpy is converted to high kinetic energy, manifested at the nozzle exit as high velocity. The nozzle is oriented to direct the high-velocity steam jet at the **blades** as indicated in Fig. 7.2, where the impact of the steam jet upon the blades develops a force sufficient to move the blade, turning the **disk** and **shaft**, as well as any load connected to the **drive flange**. Since blades are fitted uniformly around the periphery of the disk, the steam jet always has a number of blades to act upon as the disk rotates, so that the **torque** created by the impact of the steam on the blades is constant. The assembly of rotating components, including the shaft, disk, and blades, is called the **rotor**.

The **casing** serves to contain the steam and supports the **steam inlet** connection, the **steam chest**, the nozzle, and the exhaust connection. Where the shaft, which must be free to rotate, penetrates the casing, the clearance between casing and shaft is sealed by **packing glands**. The **bearings** position the rotor while allowing it to rotate: **radial bearings** support the weight of the rotor and position the shaft vertically and radially; the **thrust bearing** positions the shaft axially. The bearings are separate from the turbine casing, so that the steam leaking from the glands cannot mix with the bearing lubricating oil, or vice-versa. The bearings and casing are supported on a **bedplate**, for bolting to ship's structure via a suitable, welded foundation. Although this is a small turbine, analogous features will be found in larger turbines.

7.2 Number of nozzles

A small, low-power steam turbine may have only a single nozzle, but for more torque and power additional nozzles would be fitted in the steam chest, adjacent to the first and facing the same blades, as shown in Fig. 7.3, a transverse section through a simple steam chest. Each additional nozzle allows an increased mass of steam to be formed into an additional jet, to act on the same blades, as the blades rotate away from the preceding jet. There may thus be not one nozzle, but a row of nozzles, supplying steam to the row of blades. In the extreme case, nozzles are fitted around the full periphery of the turbine.

Fig. 7.4 shows how a steam chest may be partitioned to divide the number of nozzles into **nozzle groups**, each controlled by a valve. This arrangement enables the operator to match the number of nozzles in use to the flow of steam required to meet the power required from the turbine.

7.3 Multi-stage turbines; pressure compounding

A **turbine stage** is defined as a nozzle, or a row of nozzles, followed by one or more rows of blades. The turbine described above has only a single row of nozzles, and is therefore a **single-stage turbine**.

There is a practical limit to the amount of energy any turbine stage can efficiently convert to power. Therefore, where steam is available at high pressure and temperature (and therefore with high enthalpy) and where efficiency is important, the energy available is converted to power in steps, by arranging a series of stages on a single shaft as in Fig. 7.5. Each stage then receives its steam at the exhaust pressure of the previous stage. The flow in this turbine is from right to left, and there are seven stages. This turbine would be referred to as **pressure-compounded, multi-stage turbine**.

The turbine of Fig. 7.5 has many features in common with the single-stage turbine of Fig. 7.1: a rotor comprised of a shaft carrying a disk and blades for each stage, supported by journal bearings and positioned by a thrust bearing, rotating in a casing with steam inlet and exhaust connections, with packing glands to seal the casing where the shaft penetrates. The significant difference is in the mounting of the nozzles: although the first-stage nozzles are fitted into a plate forming the exit from the steam chest, just as in the single-stage turbine, the nozzles of the downstream stages are fitted into **diaphragms** (see also Fig. 9.8), each attached to the casing at its periphery, which partition the casing into separate chambers, one for each stage. So that steam cannot leak from stage to stage along the shaft, thereby bypassing the nozzles, each diaphragm is sealed against the shaft at its inner diameter by **interstage packing**.

As the steam flows from stage to stage through a pressure-compounded turbine, its pressure drops, of course, and its volume increases. To accommodate this **expansion** in the volume of the steam, the heights of the nozzles and blades increase in successive stages, as may be seen in Fig. 7.5.

7.4 Types of turbine stages

Three types of turbine stage are common:

Simple impulse stage. This is the only type of stage discussed and illustrated up to this point. In an impulse stage, all of the conversion of enthalpy to kinetic energy occurs in the nozzle, and all of the force developed on the blade results from the impact of the steam jet. Because no further conversion to kinetic energy of the steam occurs after it leaves the nozzle, the pressure drop is confined to the nozzle, and the pressure is then constant as the steam flows through the blades to the exhaust. Fig. 7.6 illustrates the relation between pressure and velocity of the steam in a single, simple impulse stage. Simple impulse stages are also called **Rateau stages**.

Velocity-compounded impulse stage. This type of stage is very commonly used in small turbines, and is often used as the first stage of larger turbines. See Figs. 7.7 and 7.8. In keeping with the definition of an impulse stage, all of the conversion of enthalpy to kinetic energy occurs in the one row of nozzles, but the conversion of this kinetic energy is then shared by two rows of blades mounted on the same disk. Because the direction of steam leaving the first row of blades is opposite to the direction necessary to drive the second row, as may be seen from the section of Fig. 7.7, a set of **vanes**, or stationary blades, is interposed between the two moving rows, fixed to the casing. These vanes serve simply to change the direction of the steam, directing it to the second row of blades. These vanes are not nozzles, as they do not increase the velocity of the steam. Velocity-compounded impulse stages are more commonly called **Curtis stages**.

Reaction stage. See Figs. 7.9 and 7.10. In a reaction stage, only a portion of the enthalpy of the steam is converted to kinetic energy in the nozzle, with the balance of the conversion occurring as the steam passes through the blades, where the passages between adjacent blades are so shaped by the blade profiles as to give each passage the shape of a nozzle. As the steam thus accelerated in the blades is expelled, a reaction force adds to the impact of the steam on the blades. Because the enthalpy of the steam is converted to kinetic energy in both nozzles and blades, the pressure drop is also shared by the blades, as shown in Fig. 7.10. It should be noted that the velocity increase of the steam that occurs in the blade passage is an increase in velocity only relative to the blades, while the velocity plotted in Fig. 7.10 is the absolute velocity, i.e., the vector sum of the blade velocity (acting downward in the figure) and the steam velocity (acting upward). Reaction stages are often referred to as **Parsons stages**.

The blade sections shown in the figures are characteristic of the two types, almost symmetric for impulse blading and airfoil-shaped for reaction blading.

Each type of stage has its applications. An important point to note is that in a reaction stage the pressure difference across the moving blades leads to complications not present with impulse stages. One of these complications is that except for turbines operating at very low pressures, nozzles for reaction stages must be evenly distributed around the full periphery of the turbine.

7.5 Available energy: vacuum exhaust

It was already demonstrated that it is desirable to supply the steam to the turbine at high pressure and temperature, and therefore, at high energy (or enthalpy) levels, so that the amount of energy available from the steam for conversion to power output in the turbine is high. It was noted earlier that steam generated at about 1000 psi and superheated to 955°F has an enthalpy of 1480 Btu/lbm. If the steam were to exhaust from the turbine at 1000 Btu/lbm, a typical value for propulsion turbines, each pound of steam would have yielded 480 Btu of enthalpy for conversion to power in the turbine.

However, if the steam were allowed to exhaust at atmospheric pressure, 14.7 psi, where its enthalpy is about 1160 Btu/lbm, it would yield only 320 Btu from each pound of steam. To extract more energy from the steam requires that the exhaust enthalpy, as determined by its pressure and temperature, be reduced. This reduction can be achieved only if the exhaust pressure is reduced to a vacuum: for marine propulsion turbines, a typical exhaust pressure is 1.5 inches of mercury, about 0.74 psi, at which the saturation temperature of the steam is 91.7°F, and its enthalpy is the desired 1000 Btu/lbm.

The reason for vacuum exhaust, therefore, is to increase the energy available from the steam for conversion to power in the turbine. Lower steam flow rates and higher cycle efficiencies result.

To maintain such a high vacuum in the condenser is a serious complication. However, the increase in available energy and cycle efficiency is justified in almost all steam power plant applications.

CHAPTER 8
STEAM TURBINES: EXAMPLES AND FEATURES

8.1 Cross-compound propulsion turbines

Most propulsion turbines are arranged as **cross-compound turbines**, illustrated in Figs. 8.1 and 8.2, with separately encased **high-pressure** and **low-pressure turbines**, each containing a series of pressure-compounded stages on a single rotor, with the high-pressure turbine exhaust flowing through the crossover pipe to the low-pressure turbine. The output of the two turbines will generally be combined by gearing to drive a single propeller.

Figs. 8.3 and 8.4 are sections through high-pressure and low-pressure impulse turbines that might be used in such a set. The direction of steam flow is from right to left in the high-pressure turbine, after which the steam exhausts through the crossover (here called a "cross under") to the low-pressure turbine, where it flows from left to right. The following features are worthy of note:

Because these are pressure-compounded impulse turbines, the **disk-and-diaphragm construction** described previously is used. In each of the high-pressure turbine disks an **equalizing hole** is visible, which ensures that there can be no pressure difference across the blade row.

In this high-pressure turbine the first stage is a simple impulse stage, but many turbines have Curtis stages as first stages. The first stage is also called the **control stage**, because it is here that the number of nozzles in use can be selected.

Astern stages are fitted to the low-pressure turbine. These stages are oriented to rotate the rotor in the opposite direction, thereby turning the ship's propeller backwards to slow the ship, bring it to a stop, or enable it to back down. In this example, the first astern stage is a Curtis stage.

The casing is fixed rigidly to its foundation only at one end, with the other end supported by a means that will allow thermal expansion (a **flexible-support plate** is visible in Fig. 8.3).

An **extraction connection** is visible in Fig. 8.3. Extraction connections allow some steam to be removed from the turbine for use elsewhere in the plant. Extractions are also referred to as **bleeds**.

In the low-pressure turbine, **moisture-elimination grooves** are visible after the sixth stage. Even when superheated steam is supplied to the high-pressure turbine, as its energy is removed in successive stages a point is reached, usually about half way through the low-pressure turbine, where the steam is **saturated**, i.e., no longer superheated. Beyond this point some of the steam will condense to form moisture droplets, which are best removed from the stream.

8.2 A high-pressure turbine with reaction stages

Fig. 8.5 shows a high-pressure turbine with reaction stages, of a type that might be used in a cross-compound propulsion set, as an alternative to the high-pressure impulse turbine described above. The following points should be noted, all of which result from the fact that in reaction stages there is a pressure difference across the moving blades:

The control stage is a Curtis stage. It would never be a reaction stage like the stages that follow it, since reaction stages preclude control by selection of the number of nozzles.

In contrast to the disk-and-diaphragm construction of impulse turbines, the rotor is shaped as a drum, as shown in Fig. 8.6, with the moving blades inserted into its periphery; there are no disks except for the control stage. **Drum rotors** are characteristic of reaction turbines.

Nor are there diaphragms in a reaction turbine; instead, the nozzles are formed by the passages between stationary blades, fixed to the inside of the casing, or as in this example, to the inside of stationary **blade rings** attached in turn to the casing.

Interstage packing is fitted to prevent steam from bypassing the nozzle passages, as in an impulse turbine, but in addition, because of the pressure difference across the moving blades, seals must be fitted to prevent leakage of steam over the tips of these blades, as well. See Fig. 9.10 for a detail of these **tip-sealing** arrangements.

The control stage disk, being larger in diameter than the drum of the rotor, provides an area for steam pressure to develop a force on the rotor acting to the right, as in Fig. 8.6, to balance the opposite force developed as steam flows through the reaction blades. The first stage disk therefore acts as a **dummy piston**.

8.3 A double-flow, low-pressure turbine

Fig. 8.7 shows a **double-flow**, low-pressure turbine, in which steam enters at the center, then divides into two equal streams, which flow to the right and left through identical, mirror-imaged groups of stages. Double-flow turbines are found only in very high-powered plants, where the volume of steam would demand a low-pressure turbine of excessive diameter and blade height. By dividing the flow, these very large volumes of steam can be accommodated in a turbine of modest diameter and blade height. This arrangement can be advantageous even though the turbine is longer, comprising, in effect two turbines in the same casing. Although this is a reaction turbine, the double-flow principle is equally applicable to impulse turbines.

Several features are worthy of note:

Because this is a reaction turbine, drum-rotor construction is used.

The axial forces on the rotor, developed as the steam flows over the reaction blading, will cancel, eliminating the need for a dummy piston.

In this example, the astern stages are fitted symmetrically at each end of the rotor, but they could also be grouped together at one end, as a single astern turbine.

8.4 A single-cased propulsion turbine

Although the vast majority of propulsion turbines are of the cross-compound configuration discussed above, there are some applications better served if an adequate number of stages can be confined to a single casing, as in Fig. 8.8, a **single-cased turbine**. One such application is **turbo-electric drive**, where the single turbine shaft can be directly connected to drive a single propulsion generator. Another is in nuclear submarines, as in Fig. 8.9, where a pair of single-casing turbines are geared together to drive a single propeller, providing a greater measure of redundancy than a cross-compound set.

The following features shown in Fig. 8.8 should be noted:

This is an impulse turbine, as evidenced by the disk-and-diaphragm construction. The control stage is a Curtis stage.

Astern stages are present, evidence that this turbine will be geared to the propeller. (With electric drive, the propulsion motor driving the propeller can be reversed electrically, without requiring reversed rotation of the generator.)

8.5 A ship's-service generator turbine

Fig. 8.10 shows a **turbogenerator** of the type commonly fitted to steam ships as the principal source of electricity. (Similar units are often found on motorships, where the steam is generated by the exhaust of the propulsion engine as it passes through a waste-heat boiler.) While the power produced by a set of propulsion turbines may range from below 10,000 kW to about 50,000 kW, the power required for ship's-service generators may range from about 500 kW to 2500 kW. One result is that the turbines for these generators usually have fewer stages than would be acceptable for a propulsion turbine. Fig. 7.5 is a section through a ship's-service generator turbine.

8.6 Turbines for pumps and fans

Several applications of single-stage turbines are illustrated in Figs. 8.11, 8.12 and 8.13. Single-stage turbines are adequate for many applications requiring less than 1000 kW. In addition, where low efficiency is acceptable, as in cargo pump turbines (which are used only when discharging cargo), higher power levels may be involved. Almost all steam ships have turbine-driven feed pumps, and the configuration shown in Fig. 8.11 is typical. Forced-draft fans are generally driven by steam turbines in naval vessels (in merchant ships forced-draft fans are almost always driven by electric motors). Fig. 8.12 shows a typical configuration, and also serves as an example of a vertical-shaft turbine, which is a common arrangement for small turbines.

CHAPTER 9
STEAM TURBINES: DETAILS

9.1 Turbine casings, bearing housings, and bedplates

Turbine casings are designed for disassembly, usually about the horizontal plane, as shown in Fig. 9.1. The casing halves are secured in service by the **bolted flange**, as in Fig. 9.2. Casing internals, including diaphragms, are split at the same plane, and remain attached to their upper or lower casing halves when the turbine is opened. The casing is usually steel, using all cast steel components, or, where appropriate, a mixture of castings and plate, welded together. Some auxiliary turbines have cast iron casings.

Bearing housings are also split for disassembly, and the usual practice is to fabricate the housing lower halves as parts of the lower half of the casing, either as integral components, or attached to it via bolted flanges. See Fig. 8.7. The bearing housing upper halves, however, are generally separate, in order to maintain the functional separation mentioned previously.

To attach the turbine to its foundation, brackets or flanges (called **feet**) are secured to the assembled lower half, generally in way of the bearings, so that the lower half of the casing then serves as a bedplate as well.

I o permit thermal expansion of the casing, one end is secured rigidly to the foundation in all three planes, but the other end is free to move axially, while being constrained vertically and transversely. The flexible-support plate mentioned previously is one means of allowing this expansion. An alternative is a **sliding foot**, most easily formed by elongation of the bolt holes in the flange, which may then move axially as the casing expands.

9.2 Rotors

Modern practice for large marine turbines calls for the major components of the rotor, except for the blades, to be machined in one piece from a single steel forging. This is illustrated well in Figs. 8.5 and 8.7, as well as by Fig. 9.3, which illustrates the manufacturing steps involved. Only small auxiliary turbines are likely to have rotors built up from separate components.

9.3 Blades

Most turbine blades are of corrosion-resisting steel, ground to shape from bar stock or forged blanks, with an integral **root** section and, often, an integral **tenon** for the **shroud**. See Fig. 9.4. The blades are secured in the rotor in grooves machined in the rotor to a shape complementing that of the root. They may be designed for insertion axially or circumferentially, as in Figs. 9.5 and 9.6. The complex design of the root is necessary because of the high centrifugal forces involved.

Generally, the tips of adjacent blades are attached to each other by a shroud, as in Fig. 9.7, usually in groups of six or eight, mostly as a means of stiffening the blades against vibration.

9.4 Nozzles and diaphragms

Diaphragms of impulse turbines are fabricated of welded components, as in Fig. 9.8. They are split horizontally so that each half may remain attached to the casing when the turbine is opened for inspection.

The nozzles of reaction turbines are formed by stationary vanes, usually manufactured individually like the blades, but with a simpler root, for securing in the casing, as in Fig. 8.7, or, as in Fig. 8.5, in a separate blade ring attached to the casing.

9.5 Packing glands

Most large turbines use **labyrinth packing**, illustrated in Fig. 9.9, a section through the high-pressure end of a high-pressure turbine, such as that of Fig. 8.3. The inward-pointing, circumferential ridges on the **packing rings** are not actually in contact with the shaft, but are separated by a small radial clearance. The packing does not, therefore, completely preclude leakage, but rather limits it to a tolerable amount. (Contact packing, which could stop the leakage, would, at the high shaft speeds of turbines, have unacceptable friction and wear.) As steam is forced out of the casing by the pressure within, each ridge presents an obstacle to limit the flow of leaking steam and forces its pressure to drop. After the first rings, the pressure of the leaking steam is sufficiently low to allow some of it it to flow to a lower pressure turbine stage, so that it is recovered. A similar recovery, in the **gland-seal system**, occurs after the next ring. The final recovery is to a vacuum source, the **gland leak-off condenser**, which will draw not only leaking steam but also inward leaking air. As an examination of these figures will show, at the lower-pressure end glands of propulsion turbines, and in turbogenerator turbines, the arrangement is similar, but with the first rings and recovery path deleted.

A single ring of labyrinth packing is used as interstage packing in the diaphragms of impulse turbines, as in Figs. 7.5, 8.3, 8.4, and 8.8.

The same principle, of limited leakage past a non-contact packing ring, is used for **tip sealing** of reaction blades, as shown in detail in Fig. 9.10, and also in Figs. 8.5 and 8.7.

9.6 Journal bearings

A self-aligning **journal bearing** of the type frequently used in large marine turbines is shown in Fig. 9.11. The spinning of the shaft forces the oil between the shaft surface (the **journal**) and the stationary lining of the **bearing shell**, so that the journal floats on a thin film of oil, without metal-to-metal contact, thus minimizing wear and friction. **Babbitt** is a soft metal alloy, which is cast into the shell. The spherical seating of the shell allows it to align itself with the journal even if slight distortion has occurred. The lubricating oil is circulated by an external pump, from bottom to top, leaving over the **sight glass** and thermometer, to be cooled in an external cooler. The **oil-deflector rings** help to maintain the segregation of oil and steam.

Most small turbines use journal bearings that are similar, but without the self-aligning feature. Only occasionally are anti-friction bearings (such as ball or roller bearings) used in marine turbines.

9.7 Thrust bearings

A thrust bearing of the type used in all but small auxiliary turbines is shown in Figs. 9.12 and 9.13. Bearings of this type are called **Kingsbury** or **Michell bearings**, and are classed as **pivoted-shoe thrust bearings**. The figure shows the usual case of a bearing arranged to absorb thrust in either direction, i.e., to the right or to the left, transmitted from the spinning **thrust collar** on the shaft through the **shoes** to the **base ring**, which is rigidly supported in the bearing housing, in the manner most clearly seen in Fig. 7.5. The pivots on the shoe supports, and the **leveling plates**, permit the shoes to tilt with respect to the face of the collar. By tilting the shoes a thin film of oil can be maintained between the collar and the babbitt lining of the shoe, so that the thrust is transmitted without metal-to-metal contact, and with minimum friction and wear. The lubricating oil is circulated through the thrust bearing housing by the same pump that circulates the journal bearings.

CHAPTER 10
STEAM TURBINE OPERATIONS

The procedures described here are for a steam turbine geared to a fixed-pitch propeller, but are otherwise general, intended to provide background information only, and are not comprehensive or all-inclusive. Recommended procedures for specific equipment should be sought out and followed.

10.1 Preparation, raising vacuum, warming the turbine

If a turbine has been shut down for an extended period, or if the casings of the turbines or reduction gear have been opened, verification should be sought that all foreign objects, including tools and rags, were removed before the casings were closed. All hatches and inspection openings should be closed and sealed. Rotor axial-position indicators should be checked.

Lubricating oil (LO) should be at a temperature of at least 90°F (about 30°C), clean, and in acceptable condition. The oil should be circulated through the purifier for about twelve hours prior to its use to ensure that it is clean, free of water, and to bring it up to minimum temperature. Oil filters should be cleaned. If there is any doubt about the quality of the oil, a sample should be taken for testing, and if necessary, the oil should be transferred to the LO settling tank, and the system should be drained, flushed and refilled.

The LO pump is started. When the gravity tank is filled to its normal level, the sump level is checked and corrected. Oil flow to all bearings and gear meshes is verified.

Casing and throttle drains are opened. Permission to rotate the propeller using the turning gear must be obtained from the bridge. The turning gear is engaged and started. A round is made, listening for unusual noises.

The circulating water pump is lined up and started, using the high sea suction if in port or in shallow water. Flow through the condenser and LO cooler is verified.

The condensate-recirculating valve is opened, and the condensate pump is started. With the turbine rotors rotating on the turning gear, steam is admitted to the gland seals. The gland exhaust fan or ejector is started. The air ejector is started. If a steam-jet air ejector is fitted, only the second stage is started at this point, to raise vacuum to about 15 inches of mercury (about 0.5 bar).

The main steam stop valves are checked to see that they are closed, and then the ahead and astern throttle valves are opened fully, then closed, to check their operation. Steam is then brought up to the throttle valves, with the lines drained and the drains closed. The astern steam guarding valve is opened. Permission to rotate the propeller with steam must be obtained from the bridge. The turning gear is disengaged, and immediately, the astern throttle is opened enough to start the rotors turning, then throttled down to keep the shaft turning at about five rpm, for about fifteen seconds, then closed. The ahead throttle is then opened, wide enough to start the rotors spinning, then throttled down to a shaft speed of about five rpm, and closed after about fifteen seconds. This alternating rotation is continued for about fifteen minutes. Rounds should be made while the rotors are spinning to listen for unusual noises. Turbine alarms and trips are tested.

After fifteen minutes of continuously spinning the rotors alternately ahead and astern, the throttles may be left closed for three to five minutes at a time, then spun briefly in each direction. When machinery is arranged for bridge control, an automatic rotor-spinning program is included, and the turbines can therefore be placed on bridge control at this point. In any case, this procedure is continued until maneuvering is begun.

Just prior to the start of the maneuvering period, condenser vacuum is brought up to normal.

10.2 Maneuvering

Throughout the maneuvering period, the astern guarding valve is open, the throttle and casing drains are open, and the extraction valves are closed. If the feed-pump recirculating valve is manually controlled, it is kept open throughout. The condensate-recirculating valve must be adjusted to keep a level in the drain well at all times. Both of these recirculating functions are automated for bridge control. On ships with scoop injection, the circulating pump is run throughout the maneuvering period. In some plants, steam pressure and superheat temperature are reduced for maneuvering.

During the maneuvering period, the ahead and astern throttle valves are used to meet commands, either by direct manual control, or under remote control from the engine control station or from the bridge. The throttles are never open simultaneously: whichever throttle is opened must be closed completely before opening the other. When changing speed, a throttle is opened or closed further than necessary to quickly achieve the required rpm, then adjusted to maintain the rpm. Under manual control, ahead throttle settings are simplified by using the first-stage exhaust pressure, which correlates consistently with rpm.

If there is a critical-speed range, it is passed through quickly.

When slowing down, if the way on the ship turns the propeller at a higher speed than ordered, astern steam is used to correct the rpm while moving ahead, and ahead steam is used when moving astern. Such correction is especially required when a stop is ordered, to keep the propeller from turning. During an extended stop, the rotors are periodically spun ahead and astern. These procedures are automated for bridge control.

During the maneuvering period, instrumentation is monitored, manually or automatically. The boilers, in particular, are watched closely as they respond to changes in steam demand. Depending on the extent and condition of automation present, it may be necessary to make adjustments to keep temperatures, pressures, and levels within bounds. A common precaution is to keep the standby feed pump ready, warmed, and with only its steam stop valve closed.

To help ensure continuity of electric supply, it is a common practice to keep an additional ship's-service generator on the line. Where motor-driven thrusters are fitted, the bus may be divided to isolate the thrusters from other services.

At the end of a departure maneuvering period, throttle and casing drains are closed, and the astern guarding valve is closed. The correct number of ahead nozzles for the required speed are opened, and the ahead throttle is gradually opened fully. If steam pressure and/or temperature were reduced for the maneuvering period, normal values are gradually restored. Feed-pump and condensate-recirculating valves are closed. Extraction valves are opened. Live steam to the gland-seal system is secured. If water depth permits, the circulating pump is switched to the low sea suction. On ships with scoop injection, the scoop valve is opened and the circulating pump is stopped. Additional generators are secured. At the boilers, atomizer-tip sizes and the number of burners in use are adjusted to match the steam demand.

10.3 Routine operation

While under way at steady power, all instrumentation is monitored, manually or automatically, and adjustments are made as required. Even in plants with extensive automation, engineers make regular rounds, listening for unusual noises, looking for leaks and sources of unusual vibration, and checking sight glasses and fluid levels. Trends are noted for their value in planning maintenance and in avoiding unexpected failure.

When a speed reduction is ordered, the throttle valve may be used for an immediate response, but a longer-term adjustment is made by adjusting the number of nozzles, then opening the throttle again. Nozzle groups in use should always be adjacent to each other.

The LO purifier is kept in continuous bypass service on the main sump. The purifier and LO filters are cleaned weekly, or as experience dictates, and accumulations are examined for metallic traces. Oil samples should be withdrawn monthly for an analysis of the physical and chemical properties of the oil, which can reveal incipient problems.

10.4 Securing

At the end of an arrival maneuvering period the extraction valves are closed, the astern guarding valve is open, the throttle and casing drains are open, live-steam supply to the glands is open, the feed-pump recirculating valve and the condensate-recirculating valve are open, and the turbines are being spun periodically with steam. This situation is maintained until the bridge has given and confirmed the order, "finished with engines." The main steam stop valves are then closed, and the throttle valves are used to dump any steam remaining in the lines. The throttle valves are then closed tightly, and immediately, the turning gear is engaged and started.

Air ejectors are secured, but gland-sealing steam is maintained, so that vacuum breaks though the air-ejector vent, rather than through the packing glands. When the condenser reaches atmospheric pressure, gland-sealing steam supply is closed, and the gland exhaust fan or ejector is stopped. When the condenser drain-well level is normal, the condensate pump and circulating pump are stopped.

After several hours, when the turbine rotors have cooled to ambient temperatures, the turning gear is stopped (but not disengaged) and the LO pump is stopped.

Every other day at idle, a LO pump is started, and when oil flow is confirmed at the bearings and gears, the turning gear is started and run for about fifteen minutes, taking care to stop the shaft in a different position each time. At the same time, the air ejector is run to draw fresh air through the turbines. With steam-jet air ejectors, the second stage is sufficient for this purpose. If possible, the LO purifier should be run continuously on the main sump, with steam to the heater, to prevent condensation within the gear case.

10.5 Emergency operation

Cross-compound turbines are required to be fitted with arrangements for bypassing either turbine, in case of a problem with the turbine or its gear train. Generally, with one turbine, less than half of the rated power can be achieved because of torque limits on shafts and gears. A typical arrangement involves blanked flanges at the throttle, crossover, and condenser inlet, the necessary sections of connecting piping, and orifices to limit steam flow. If the high-pressure turbine is to be bypassed, the ahead throttle is connected directly to the crossover by the prefabricated emergency pipe section, with an orifice inserted after the throttle valve, and a blank is inserted in the high-pressure exhaust flange. The high pressure turbine is disconnected from the gearing. In service, the ahead throttle is used to control steam pressure, up to a maximum set by the orifice. It may be necessary to reduce the steam temperature, as well. If the astern turbine is on the low-pressure rotor, it remains available.

If the low-pressure turbine is bypassed, the crossover is connected directly to the condenser inlet, and a blank is inserted at the low-pressure turbine inlet. An orifice is inserted after the ahead throttle valve. The bypassed low-pressure turbine is disconnected from the gearing. In operation, the ahead throttle valve is then used to control steam pressure, up to the limit set by the orifice. If the astern turbine is on the low-pressure rotor, no astern power is available. In fact, the astern steam line should be blanked off, or the astern valves wired closed as a safety precaution.

If, after a prolonged stop, a rotor vibrates or squeals as the throttle is opened, it may be because the rotor is warped from uneven heating. The problem can often be solved with minimal damage by spinning the rotor at low speed for several minutes to heat it evenly.

If lubricating oil pressure fails while the shaft is turning, the throttle will close automatically. However, with way on the ship, the shaft will continue to be turned by the propeller. Consequently, the shaft must be stopped, and held at a stop, using steam, although permission to do so must be obtained from the bridge. The emergency supply of lubricating oil in the gravity tanks will not be sufficient for more than a few minutes of operation. If permission cannot be granted, for example, during an emergency full astern, it is because the risk of bearing and gear damage is an acceptable trade-off against greater danger to the ship or to personnel.

10.6 Starting auxiliary turbines

A generalized procedure is described here. The steps actually required depend on the particular installation and application.

The unit is first inspected to ensure that it is closed up and ready for service. A small unit might be turned over by hand while the operator listens for any scraping noise or any sign of resistance. Oil levels are checked. If there is a pressure-fed lubrication system the priming pump is operated. Availability of cooling water, if required, is verified. Linkages for governor valves or manual trips are inspected for freedom of operation. Steam drains are opened.

It is necessary to have the rotor spinning soon after steam is admitted so that it warms evenly. A noncondensing turbine is started by opening the exhaust valve fully, then opening the steam inlet valve sufficiently to start the turbine turning. For a condensing turbine exhausting to an independent condenser, the circulating water flow is established, then the steam inlet valve is opened sufficiently to start the turbine turning; if there is a gland-seal steam supply it is then opened, and vacuum is raised. For a turbine exhausting to a condenser already under a vacuum, the exhaust valve is opened slightly, then the steam inlet valve is opened sufficiently to start the turbine turning; if there is a gland-seal steam supply it is then opened, and then the exhaust valve is opened completely. Once the turbine is turning and the exhaust is established, the drain valves are closed and the steam inlet valve is opened until the governor or other control device takes charge.

It is important to periodically test the safety devices, especially the overspeed trip, and it is common to do this when starting the turbine, but before placing it under load. After the devices are proven, they are reset and the steam inlet valve is again opened until the governor or other control device takes charge. At that point the turbine is ready for load.

CHAPTER 11
DIESEL ENGINES: BASIC PRINCIPLES

11.1 A two-stroke cycle, low-speed engine

Fig. 11.1 shows a diesel engine of the type often used for main propulsion of merchant ships. This particular example has six cylinders, a fact which is evident in the figure. Of the six cylinders, counting from the left, cylinders 1 and 2 are shown intact, while 3, 4, and 5 are cut away. The **bore**, or cylinder internal diameter, of this engine is 600 mm, and overall, the engine measures some nine meters in length, ten meters in height, and weighs almost 400 tons. At its **maximum continuous rating (MCR)** the engine can deliver some 12,000 kW, while operating at a crankshaft speed of 110 rpm, a relatively low speed that is suitable for propeller drive without any gearing or other transmission. Larger engines of this same general configuration, with as many as twelve cylinders, can develop over 50,000 kW, and weigh over 1000 tons. Their large dimensions and weight limit them to merchant ship types, where their efficient use of poor quality fuel oil and the simplicity of **direct drive** of the propeller are advantageous.

Because of its simplicity this **two-stroke, low-speed engine** will be used to introduce nomenclature and basic principles before moving on, in the next chapter, to introduce **four-stroke cycles** and **medium-speed** and **high-speed engines**.

11.2 Major components; piston motion vs. crankshaft torque

Fig. 11.2 is a schematic representation of cylinder 3 in Fig. 11.1, viewed from the aft end of the engine, which is at the right side of Fig. 11.1. Take a moment to relate the labeled items in Fig. 11.2 to their actual shapes in Fig. 11.1. Identify the **cylinder head**, with the **exhaust valve** and **fuel injector**; the **cylinder** and its **liner**, or inside surface; the **piston** and its **piston rings**; the **piston rod**; the **crosshead** and **crosshead pin**, **crosshead guides**; **connecting rod**; and **crankpin** (hidden in Fig. 11.1), **crank**, and **crankshaft journal**.

If the engine were running, turning the crankshaft clockwise, the piston in cylinder 3 would be moving down in the cylinder, forced down by the pressure of expanding gas in the combustion chamber. This downward motion is transmitted by the piston rod to the crosshead, and from the crosshead, through the connecting rod to the crank of the crankshaft. While the crosshead is constrained by the crosshead guides and thus can move only up and down, the connecting rod, which pivots on the crosshead pins at its upper end, converts this reciprocating motion to turn the crankpin in a circle, thereby rotating the crankshaft about the center of its journal. The vertical force of the gas in the cylinder is thereby converted to a rotating torque on the crankshaft.

Note from Fig. 11.2, that the vertical position of the piston in the cylinder relates to the angular position of the crank: in the figure, the crank is just past the three o'clock position, while the piston is about halfway down the cylinder. When the piston was at the top of its stroke, the crankpin was at twelve o'clock, positions that, for both piston and crank, are referred to as **top dead center**, or **TDC**. When the piston is at the bottom of its stroke, with the crank at six o'clock, they are at **bottom dead center**, **BDC**. The distance traveled each way by the piston, between TDC and BDC or vice-versa, is the **stroke**. Since this distance is also the vertical displacement of the crankpin between its dead center positions, the stroke is therefore twice the radius, or **throw**, of the crank, measured from center of the crankshaft journal to the center of the crankpin. Note that one complete revolution of the crankshaft requires the piston to make two strokes, one down and one up.

In Fig. 11.1, the three cut-away cylinders show pistons at different positions, determined by the angular positions of the cranks on the crankshaft, which are fixed when the crankshaft is manufactured. Thus, while the piston in cylinder 3 is moving down under pressure of gas in its combustion chamber, and is about halfway through its power stroke, the piston in cylinder 4 is slightly below its TDC position, at the beginning of its power stroke, and the piston in cylinder 5 is just past its BDC position, starting to move up. While the pistons of cylinders 3 and 4 are being pushed downward on power strokes by gas in their combustion chambers, delivering force to turn the crankshaft, the upward movement of the piston in cylinder 5 reverses the direction of force: as the crank of cylinder 5 rotates, it will push upward on the connecting rod to push the crosshead and piston upward.

11.3 Two-stroke cycle events

Note, in Fig. 11.1, the row of circular **air ports** visible in cylinders 3 and 4, just above the bottom of each cylinder. Note that, in cylinder 5, these ports have just been blocked by the upward movement of the piston. In this position, cylinder 5 of Fig. 11.1 corresponds to (a) in Fig. 11.3. Fig. 11.3 shows a complete cycle for one cylinder of this engine. In describing the cycle, reference will also be made to Fig. 11.4, which shows the pressure in the cylinder, in the space between the top of the piston and the cylinder head, plotted against the piston position. Starting with Fig. 11.3(a):

> When the piston closes the air ports (and after the exhaust valve has closed) the air, which filled the cylinder while the air ports were open, is trapped above the piston, and will be compressed as the piston rises, until its volume is less than a tenth of the original value, as in Fig. 11.3(b). As the trapped air is compressed its pressure rises as in Fig. 11.4, as does its temperature.
>
> The temperature reached by the trapped air towards the end of the **compression stroke**, as in Fig. 3(b), will be high enough to ignite fuel when the fuel is injected. Fig. 11.3(b) shows the cylinder just as **fuel injection** is about to begin. Before the piston passes through TDC the first droplets of fuel have ignited and begun to burn, raising the pressure in the cylinder, as shown in Fig. 11.4. This **combustion process**, which is well under way in Fig. 11.3(c), continues until the last of the fuel to be injected has burned, generating the high pressure gas that will force the piston through its **power stroke**.
>
> Fig. 11.3(d) shows the piston about half way through its power stroke. Fuel injection has ended (a point called **cut off**) followed by the end of the combustion process. The gas pressure in the combustion chamber is still high (see Fig. 11.4) and forces the piston through the rest of its power stroke as the gas expands.
>
> As the piston approaches BDC the exhaust valve is opened, and the gas in the cylinder rushes out to an exhaust manifold (an action called **blow down**). As the piston moves farther down it exposes the air ports, and air flows into the cylinder under pressure, pushing the remaining exhaust gas out as the piston moves through BDC, as shown in Fig. 11.3(e). This process is called **scavenging**.
>
> As the piston passes through BDC and starts up again, the cylinder is largely full of incoming air. When the piston covers the air ports, and after the exhaust valve has closed, as in Fig. 11.3(a), the air will be trapped and the next cycle will have begun.

The cycle described above requires two strokes of the piston (one up, then one down), and therefore one revolution of the crankshaft. The processes of blow down, exhaust, scavenging, and charging with fresh air, collectively referred to as **gas exchange**, occur while the piston is passing through BDC. Because each cylinder provides a power stroke for only part of one revolution, multiple cylinders are necessary, with their cycles out of phase, as set by the angular position of the cranks on the crankshaft, to generate continuous torque to drive the load.

11.4 Camshaft action in a two-stroke engine

In the two-stroke engine of Fig. 11.1 two important cycle events, fuel injection and exhaust valve operation, occur through the action of the **camshaft**. The camshaft is driven from the crankshaft by the **timing gears**. Identify these components in the figure.

The action of the camshaft in operation of the exhaust valves will be described with reference to Fig. 11.5, which shows the relevant components for an engine that is similar to the engine of Fig. 11.1. The cam comprises a circular portion, the **base circle**, and a protruding **lobe**. The **cam follower** is held against the surface of the cam by the spring, and as the cam on the camshaft rotates, the lobe forces the cam follower to rise, and then permits it to fall. In this example the cam follower motion is transmitted to the exhaust valve hydraulically: the cam follower is connected to a small piston, which reciprocates in a small oil-filled cylinder, connected by an oil-filled pipe to a similar cylinder and piston at the top of the exhaust valve stem, also filled with oil. The motion of the cam follower is thereby communicated to the exhaust valve: when the follower rises on the lobe of the cam, the exhaust valve opens by moving down into the cylinder.

In many engines, the motion of the cam follower is transmitted to the exhaust valve mechanically, through a **push rod** and **rocker arm**, as illustrated in Fig. 11.6. Confirm this concept by tracing the motion in Fig. 11.6.

Fuel injection also occurs due to camshaft action, and will be described separately.

For the cycle events to occur at the correct points in the cycle, the angular position of the cams on the camshaft, and the relation of the camshaft to the crankshaft, set by the timing gears, must be correct. Since the events controlled by the camshaft occur once each cycle for each cylinder, and since there is one crankshaft revolution per cycle of a two-stroke engine, the camshaft of a two-stroke engine runs at the same speed as the crankshaft.

In some of the newest engines the timing of cycle events is achieved using electronic or magnetic sensors to detect crankshaft position, and valve operation and fuel injection are achieved with computer-controlled electro-mechanical or hydraulic actuators. The timing gears and camshaft are then unnecessary. An important advantage of these systems is the ability to adjust the timing of the events to continuously optimize engine performance for any combination of load and operating conditions, by sending appropriate input to the computer. For example, these engines can be adjusted for low exhaust emissions in port and low fuel consumption at sea.

11.5 Ignition and combustion

What is generally considered to set the diesel engine apart from other internal combustion engines, specifically the gasoline engine, is the use of **compression ignition**, which may be described as the process in which fuel, sprayed into a mass of air heated to a high temperature by being compressed, ignites from this heat alone, without a spark. Several issues are involved:

> Only air is trapped in the cylinder and compressed during the compression stroke, enabling high compression ratios to be used and high temperatures to be reached. Since the fuel is introduced only at the very end of the compression stroke it cannot ignite prematurely. (In comparison, in gasoline engines, the fuel is already mixed with the air before it enters the cylinder, and if compressed too intensely, will ignite while the piston is still rising.)
>
> A broad range of fuels can be used in diesel engines, all of lower volatility and greater availability than gasoline. These fuels, which range from **light distillate fuel** to **heavy fuel**, are cheaper than gasoline and also safer to store and handle.

Combustion takes place over an extended portion of the cycle, in a manner controlled by the rate of injection, and in a lean atmosphere of hot excess air. These factors are conducive to complete combustion, even of fuel of inferior characteristics.

The result is that in comparison to gasoline engines, diesel engines can develop power efficiently from fuels of lower quality.

11.6 Fuel injection

In the engine of Fig. 11.1, fuel injection occurs by camshaft action. At each cylinder, a cam on the camshaft drives a **fuel injection pump** in a manner similar to that of the exhaust valve hydraulic pump described above. Some of the fuel pumps and their cams are visible in Fig. 11.1. Injectors are visible in Figs. 11.1, 11.3, and 11.5, in all cases offset in the cylinder heads to allow the exhaust valves to occupy the centers of the cylinder heads. A pipe connects the discharge of the fuel pump to the injector. Fig. 11.7 shows a fuel injection pump and injector in detail. (Additional views of injection equipment appear as Figs. 14.32, 14.33, and 14.34.)

With reference to Fig. 11.7, rotation of the **cam** forces the **cam follower** and pump **plunger** into a vertical reciprocating motion. The pump plunger is simply a small piston, and the **barrel**, its cylinder. When the pump plunger is at the bottom of its stroke, with the follower still on the base circle of the cam, fuel flows through the **fuel ports** to fill the barrel. When the pump plunger moves up it blocks the fuel ports, then raises the pressure of the fuel trapped in the barrel, and then in the discharge pipe to the injector, until the pressure is high enough to force open the injector, which is otherwise held closed by a spring. When the high-pressure fuel oil forces the **injector plunger** to rise the fuel blasts into the cylinder through the orifices in the tip of the injector. The passage of the fuel oil through these orifices at high pressure causes it to break up into small droplets, a process called **atomization**. As these droplets spray into the hot air in the cylinder they absorb heat, begin to vaporize, and ignite, thus beginning the combustion process.

When the requisite amount of fuel has been injected the fuel ports in the pump barrel are re-opened by mechanical action, causing an almost immediate drop in the pressure of the fuel oil, even as the pump plunger continues to rise. The sudden drop in pressure causes the injector to close sharply, cutting off the injection, and then ending the combustion process.

In the simple case, the timing of the mechanical action which re-opens the fuel ports, and thus **meters** the fuel oil, that is, controls the amount of fuel injected, is set by the **fuel rack** of the pump, visible in Fig. 11.7. The fuel rack is in turn positioned in response to the power demanded from the engine: to develop more power, more fuel must be injected in each cycle, and therefore the fuel ports must be re-opened later in the stroke of the pump plunger. How this is achieved is explained in detail in Chapter 14.

As noted above, in some of the newest engines fuel injection is timed and controlled electronically, without a camshaft. Some of these engines use a **common rail** fuel system, in which one or more fuel pumps discharge continuously into a high-pressure fuel manifold to which all of the injectors are connected. Each injector is opened and closed in response to signals from the computer to initiate and terminate fuel injection.

11.7 Supercharging, turbocharging, and aftercooling

The maximum amount of fuel that can be burned in a cylinder during one cycle, which determines the amount of power that can be developed by an engine at any particular rpm, is limited by the amount of oxygen, and therefore, by the amount of air, trapped in the cylinder at the start of compression. A cylinder of given dimensions can hold more air, and therefore burn more fuel and develop more power, if the air supplied to the cylinder is increased in density by compressing it and supplying it to the cylinder under pressure. This compression of the charge air is called **supercharging**. To drive the supercharger the considerable energy in the exhaust gas can be used, by directing the exhaust gas to pass through a turbine after leaving the cylinders. The combination of exhaust gas turbine and supercharger on the same shaft is a **turbocharger**, and a typical example is shown in Fig. 11.8, with parts identified in Fig. 11.9. Turbochargers are visible in Figs. 11.1 and 11.6. The relation of a turbocharger to an engine is shown schematically in Fig.11.10.

The **air cooler**, examples of which appear in Figs. 11.1, 11.6, and 11.10, enhances the effect of the turbocharger. Although the density of the charge air was increased by compressing it in the turbocharger, the compression process also raised the temperature of the air. The density can be further increased if the air is cooled after compression (hence the term **aftercooler**, or, because it is located between the turbocharger and engine, **intercooler**). The air cooler is usually circulated with engine cooling water.

While the principal purpose of turbocharging is to increase the power output of an engine of given size and weight, it also results in better scavenging and charging, better combustion, lower **specific fuel consumption**, and reduced emissions. Almost all large marine engines are turbocharged and aftercooled.

For the two-stroke engine there is an additional benefit in supercharging, since the pressure of the charge air will push it into the cylinder through the air ports as they are opened by the piston, providing the force to scavenge and then charge the cylinder. If the engine were not supercharged some means of providing charge air pressure would be necessary, often in the form of a low-pressure, **scavenge-air blower**. Even on turbocharged two-stroke engines, at low power output where the turbocharger is ineffective, a **boost blower** is necessary. An electric motor-driven boost blower is shown in Fig. 11.10.

11.8 Loop-scavenged engines

The engine of Fig. 11.1, with air ports at BDC and exhaust through a valve-controlled port in the cylinder head, is **uniflow** scavenged, meaning that the incoming scavenge air moves in only one direction, upward, to force the exhaust gas out of the cylinder, as shown in Fig. 11.3(e). An alternative arrangement uses **loop scavenging**, and is illustrated in Figs. 11.11 and 11.12. Note that the loop-scavenged cylinder has no exhaust valves, but instead, exhausts through a second row of ports in the cylinder, just above the air ports. This works well on the downstroke, where the piston first opens the exhaust ports, enabling blow down of the gas before the air enters. However, the scavenging process which follows is not so thorough as with uniflow-scavenged engines. Furthermore, on the upstroke, compression cannot begin until the piston blocks the exhaust ports and traps the charge air. Loop scavenging was popular for its simplicity and reliability, but cannot achieve the fuel efficiency of uniflow scavenging. While uniflow scavenging is now preferred for large, two-stroke engines, many loop-scavenged engines remain in service.

11.9 Summary of basic principles

The following principles and concepts have been introduced, and are essential to an understanding of diesel engines:

The conversion of the reciprocating forces on the piston to torque on the crankshaft.

The processes of compression, combustion, expansion, exhaust, and charging that comprise two-stroke cycles.

Compression ignition, in which fuel sprayed into a mass of air heated to a high temperature by being compressed, ignites from this heat alone, without a spark.

Camshaft action, by which a rotating cam can be used to produce a reciprocating motion with a short stroke, used for valve and fuel pump operation.

Hydraulic transmission of force, by which an incompressible fluid is used to transmit a force through a pipe, used in fuel injection, and in the operation of low-speed engine exhaust valves.

Supercharging, especially with air cooling, to increase the density of the charge air, increasing the amount of power developed by an engine of given dimensions, and turbocharging, in which the power for supercharging is derived from the energy in the exhaust gas.

The use of a sliding piston to open and close ports in its cylinder during its stroke, used in fuel injection pumps, and in the cylinders of two-stroke engines.

CHAPTER 12
DIESEL ENGINES: FOUR-STROKE CYCLE ENGINES

12.1 A four-stroke cycle, trunk piston, medium-speed engine

Low-speed engines, as so far discussed, are intended for direct drive of ships' propellers, and are therefore limited to low crankshaft rpm. Low-speed engines are consequently large and heavy in proportion to the power they develop. A faster-turning engine, however, can be much lighter and more compact per unit of power output. These higher-speed engines require reduction gearing when used for ship propulsion, but are suitable for direct-drive of generators.

An example of a medium-speed engine is shown in Fig. 12.1. The engine has six cylinders. Counting from the left towards the drive flange at the right, cylinders 3 through 5 are cut away. Fig. 12.2 is an enlarged cross-section of another medium-speed engine, different from that of Fig. 12.1 but with very similar features. Comparison with the low-speed engine of Fig. 11.1 will yield many similarities, but two important differences:

> There are no crossheads in the engines of Figs. 12.1 and 12.2, but instead the upper ends of the connecting rods pivot on **wrist pins** carried directly by the pistons, as is most evident in Fig. 12.2, or in cylinder 4 in Fig. 12.1. The wrist pins here have the same function as the crosshead pins of the low-speed engine, while the functions of the crosshead and crosshead guides must be assumed by the lower portions of the piston and cylinder liner. Engines with this configuration are called **trunk-piston** engines, and, while all large low-speed engines have crossheads, almost all medium- and high-speed engines are of the trunk-piston type.
>
> There are no ports in the sides of cylinder liners in the engines of Figs. 12.1 and 12.2, but instead air and exhaust enter and leave through valves in the cylinder heads. This arrangement is characteristic of the four-stroke cycle, which is described below. While all large, low-speed engines use the two-stroke cycle, as do some medium- and high-speed engines, most medium- and high-speed engines use the four-stroke cycle.

Apart from these differences, the basic principles of reciprocating piston motion, and its conversion by the connecting rod and crank to a rotating torque on the crankshaft, are as previously described: a piston descending from TDC to BDC on a power stroke will push downward on the connecting rod to rotate the crankpin in a circular arc from its 12 o'clock position to its 6 o'clock position (as in cylinders 3 or 4 in Fig. 12.1, if the engine is turning clockwise). To complete the revolution, as the crankpin moves back towards 12 o'clock, it pushes up on the connecting rod and piston, as in cylinder 5 of Fig. 12.1.

The engine of Fig. 12.1 has a bore of 200 mm, is about three meters long and under 2.5 meters high, weighs 10 tons, and can develop 960 kW, or 160 kW per cylinder, at 1000 rpm. The largest medium-speed engines are capable of over 1500 kW per cylinder at about 400 rpm.

12.2 Four-stroke cycle events

Four-stroke cycle events are described with reference to Fig. 12.3:

> The intake or **charging stroke**. The **air valve** is open but the **exhaust valve** is closed. The piston has passed through TDC and is being moved down by the connecting rod as the crankshaft rotates. As the piston descends, air flows into the cylinder. Power to turn the crankshaft is provided by the other cylinders.

The **compression stroke**. The air valve closes as the piston passes through BDC, trapping the charge of air in the cylinder. The piston is driven up as the crankshaft rotates, compressing the charge to less than one-tenth of its initial volume. As the charge is compressed its temperature rises, and toward the end of the stroke it is well above the ignition temperature of the fuel.

Fuel injection. Fuel injection begins at the end of the compression stroke, before the piston reaches TDC. **Ignition** occurs as soon as the first droplets of fuel are heated to ignition temperature by the hot air in the cylinder.

The **power stroke**. After the piston passes through TDC, the pressure developed by the combustion of the fuel begins to force the piston down. As the volume above the piston increases the continued combustion maintains the pressure in the cylinder until injection and then combustion cease (points which are called, respectively, **cut-off** and **burnout**). After burnout, the piston continues to be forced down by the expanding gas.

The **exhaust stroke**. The exhaust process actually begins just before the piston reaches BDC, when the exhaust valve opens and the residual high pressure in the cylinder is relieved as the gases **blow down**. As the crankshaft pushes the connecting rod and piston up, most of the gas remaining in the cylinder is forced out.

As the piston passes through TDC, the air valve opens before the exhaust valve closes, enabling **scavenging** of the remaining gas to occur during an **overlap** period, when both valves are open. As the exhaust valve closes, leaving only the air valve open, the cycle repeats.

The **gas exchange** process is therefore spread out over two strokes of the four-stroke cycle, enabling piston motion to assist in exhaust and charging. However, a complete four-stroke cycle requires two revolutions of the crankshaft. With only one stroke in four being a power stroke, continuous torque requires multiple cylinders.

12.3 Camshaft action in a four-stroke engine

The camshaft of a four-stroke engine operates the air valves as well as the exhaust valves and fuel injection pump. The valve action in medium- and high-speed engines is usually achieved through **push rods** and **rocker arms**, visible at cylinder 3 in Fig. 12.1. Each cylinder of this engine has two air valves and two exhaust valves; the valves visible in the figure are both exhaust valves, while the air valves are directly behind them. Each pair of valves is linked together to enable each push rod to operate them simultaneously.

In a four-stroke cycle, camshaft-controlled events occur only once per cylinder for every two revolutions of the crankshaft. Therefore, the camshaft of a four-stroke engine is driven at half the speed of the crankshaft, usually by **timing gears**. In Fig. 12.1, although the camshaft is cut away in way of cylinders 4 through 6, the camshaft gear is visible as item 16.

In some of the newest engines the timing of cycle events is achieved using electronic or magnetic sensors to detect crankshaft position, and valve operation and fuel injection are achieved with computer-controlled electro-mechanical or hydraulic actuators. The timing gears and camshaft are then unnecessary.

12.4 Fuel injection, ignition, and combustion

The methods of fuel metering and injection described previously for two-stroke engines are equally applicable for four-stroke engines, as are the descriptions of ignition and combustion. A **fuel pump** is visible in Fig. 12.1 as item 13, while an **injector** can be seen at the center of cylinder 4, as item 14.

12.5 Supercharging, turbocharging, and aftercooling

The principles of supercharging, turbocharging, and aftercooling are as valid for four-stroke engines as they are for two-stroke engines, and the earlier discussion can be reapplied. In Fig. 12.1, the **turbocharger** is item 17. The **aftercooler** is directly below the turbocharger but is obscured in this view.

Four-stroke engines do not require boost blowers or scavenge blowers, even if they are not supercharged. The downward motion of the piston on the intake stroke, with the air valves open, draws an adequate charge of air into the cylinder, while, on the exhaust stroke, with the exhaust valves open, the upward motion of the piston expels most of the exhaust gas.

12.6 Comparison of two- and four-stroke cycles

In a two-stroke engine, every cylinder completes a power stroke in every revolution, while four-stroke cylinders require two revolutions for every cylinder to complete a power stroke. It would seem, therefore, that if all else were equal, a two-stroke engine would develop twice the power of an otherwise similar four-stroke engine. However, the extra strokes of a four-stroke engine permit better cooling between power strokes, while the absence of ports in the cylinder liner removes a source of stress concentration. The power strokes of four-stroke engines can therefore be more intensive, at higher average pressures, which largely compensate for the additional strokes, at least in the medium- and high-speed engines, where the overall performance of two-stroke engines is similar to that of four-stroke engines of similar dimensions. In the large low-speed engines, fairly high average pressures can be achieved even with the two-stroke cycle, which explains the complete dominance of the two-stroke cycle for engines of that type.

CHAPTER 13
DIESEL ENGINE TYPES, CONFIGURATIONS, FUELS, AND FEATURES

13.1 Classification and categories

It should already be clear that diesel engines are **reciprocating**, **internal-combustion**, **compression-ignition engines**. Within this broad classification, it is traditional and helpful to categorize engines by other characteristics. Some of these categories are presented here.

Classification by cylinder grouping: The engines discussed in the opening chapters are **in-line engines**, as contrasted with **V-engines**, such as those on the right in Figs. 13.1 and 13.2. While all low-speed engines of current design are in-line engines, medium- and high-speed engines are built with up to eight or nine cylinders in-line, and in V configurations with as many as 24 cylinders. In the figures, note that a V-engine consists of two banks of cylinders, with as many components as possible identical to the in-line version. The banks are joined at the crankshaft, where the connecting rods of each pair of cylinders are acting on the same crank, as in Fig. 13.2. The cylinders of these V-engines are therefore not in the same transverse plane, but are offset axially. The V arrangement permits a large number of cylinders to be accommodated within a reasonable length.

Classification according to speed: Traditionally, diesel engines are grouped into categories of low, medium and high speed, depending on crankshaft rpm or mean piston speed. Low-speed engines might best be defined as those whose crankshaft speeds are a suitable match for direct connection to a ship's propeller, without reduction gearing, and so tend to have rated crankshaft speeds below 250 to 300 rpm. The medium-speed group starts at about 400 rpm. Most engineers would place the upper limit of the medium-speed group, and the start of the high-speed group, in the range of 900 to 1200 rpm. With reference to other characteristics, low-speed engines are usually two-stroke, in-line, crosshead engines, while medium- and high-speed engines may be two- or four-stroke, in-line or V, and are almost always trunk-piston types.

Classification according to operating cycle, two-stroke or four-stroke, as previously discussed.

Classification according to air supply: As previously discussed, most marine engines, especially the larger ones, are turbocharged and aftercooled. Engines which are not supercharged are said to be **naturally aspirated** (or normally aspirated). This terminology is difficult to apply to two-stroke engines that are not turbocharged, since they must be provided with some means of supplying air to the cylinders, such as a low-pressure scavenge blower.

Classification according to running gear: The term **running gear** refers to the crosshead, connecting rod, and crankshaft assembly, or, in trunk-piston engines, the piston, connecting rod, and crankshaft assembly. In crosshead engines the piston is guided in the cylinder by the crosshead and guides, which assume the sideways forces that result from the connecting rod and crank motion. In trunk-piston engines, the cylinder liner itself serves to guide the piston and must therefore carry the sideways forces.

Classification by direction of gas force on the piston: All engines in current use are **single acting**, in which combustion acts on the top of the piston, forcing it toward the crankshaft, versus **double acting**, in which combustion occurs alternately on top and bottom of the piston, producing power thrusts alternating toward and away from the crankshaft. Some engines have **opposed pistons**, as in Fig. 13.3, in which combustion takes place between two pistons in each cylinder, each of which is single acting.

Classification according to cylinder proportions: Cylinder proportions may be expressed as the **stroke-to-bore ratio**. Low-speed engines may have very high ratios of 3:1 or more, but medium- and high-speed engines have ratios close to one.

Classification according to cooling: Most engines are water cooled, i.e., water is circulated through cooling passages around the combustion chamber. In most cases, marine engines are water cooled in a closed circuit by treated freshwater, which is then cooled in a closed heat exchanger by seawater. (In some applications, such as emergency generator engines, the heat exchanger may be an air-cooled radiator as in automotive applications.) In any event, the lubricating oil serves as an intermediate coolant of the bearings, and in most cases, of the piston as well. Air-cooled engines, in which air is circulated over the external surfaces of the engine, are only occasionally encountered in marine service.

13.2 Fuels for diesel engines

Fuels used in marine diesel engines cover a spectrum from the refined **distillate fuels** called **gas oil** or **diesel fuel**, through to **heavy fuels**, which are principally residuals of the refining process blended with distillate fuels. The large, low-speed engines and many of the medium-speed engines are capable of operation on the heaviest grades of fuel, and for these engines, fuel selection is mostly an economic decision, requiring that a balance be struck between the lower cost of the heavier fuel oils, and the inconvenience and greater cost of the fuel treatment and increased engine maintenance associated with their use. Some medium-speed engines and most high-speed engines require distillate fuels.

The propulsion engines of most sea-going merchant ships are run on heavy fuel. In the past, the generator engines of these ships would be run on distillate fuel, but increasingly these are also run on heavy fuel. Most coastal and harbor craft, fishing boats, and naval vessels of all types, use distillate fuel for main and auxiliary engines.

Fuel cleanliness is most important, as it will have a direct effect on the wear of engine components that come into contact with the fuel or its combustion products. Consequently, in all but the smallest plants, fuels are treated by settling, centrifuging, and filtering. Distillate fuels have low viscosity and are generally handled and delivered to the engine at ambient temperature, but the high viscosity and specific gravity of heavier fuels both affect these treatment processes adversely, so that the heavier the fuel, the more complex will be the fuel supply system. Viscosity of fuel alone may present no problem as long as the fuel can be heated sufficiently at each point in the system to permit pumping, settling, filtration, centrifuging and atomization. The last point is also of high importance, with the heaviest fuels requiring heating to 160°C.

The **ignition quality** of a fuel is an indication of the time necessary for the fuel to ignite after it has been injected into the cylinder of an engine. The ignition quality of distillate fuels is generally adequate for even the highest-speed engines, but the low ignition quality of some heavy fuels may be troublesome for even the low-speed engines.

The **carbon residue** index of a fuel indicates its tendency to leave deposits after combustion, fouling and wearing cylinders, piston rings, ring grooves, exhaust valves and turbocharger turbine nozzles, and fouling injector tips, tips, interfering with injection and combustion.

Sulfur is present to some extent in all commonly available fuels, including distillates, but there may be ten times the concentration of sulfur in even the lighter blends of heavy fuel. The sulfur leads to the formation of gaseous sulfuric acid during combustion. At temperatures below about 150°C, condensation of the sulfuric acid begins, leading to acidic attack of exposed surfaces, called **cold-end corrosion**, and to acid contamination of the lubricating oil. Among the engine components which are most vulnerable to cold-end corrosion are the cylinder liners themselves, if they are cooled excessively at low power.

Vanadium is present in most heavy fuels but is rarely of significance in distillate fuel. During the combustion process gaseous vanadium oxides will form, some of which will form adhering deposits on combustion space surfaces whose temperatures exceed about 500°C. The surfaces most susceptible to such deposits are the exhaust valves of medium-speed engines operated on high-vanadium fuels, which can overheat and burn as a result.

13.3 Starting and reversing

Starting a diesel engine involves turning the crankshaft until a charge of air in one cylinder undergoes a compression stroke, raising its temperature so that an injection of fuel will ignite, at which point the engine will accelerate to idle speed. Some engines, mostly smaller engines, are started by motors that engage gear teeth on the engine flywheel. These motors may be electric motors run from batteries, or pneumatic motors operated from stored compressed air, or hydraulic motors driven by oil or water stored under pressure. Larger engines, including most propulsion engines, are started by introducing compressed air directly into a cylinder whose piston is at the beginning of its power stroke, through a starting-air valve in the cylinder head. The compressed air will force the piston down, rotating the crankshaft through almost half of a revolution. After this action is repeated at one or two more cylinders in succession the required compression stroke will be achieved, and fuel can be injected, and the engine will start.

Many of these large propulsion engines are **direct-reversing engines**, capable of being run in either direction. If compressed air is admitted to a cylinder whose piston was below TDC but approaching the end of a compression stroke when the engine stopped, the air will push the piston down, reversing the direction of crankshaft rotation from what it was before the engine stopped. For the engine to start and run in the reversed direction, however, valve operation and fuel injection must be made to occur in the right sequence. In four-stroke engines a second set of cams is necessary, usually on the same camshaft as the ahead cams, but axially displaced, and brought into position by moving the camshaft before starting the engine in reverse (see Fig. 13.4). In two-stroke engines only a slight shift in timing is required, and it is often sufficient to change the angular position of the cam followers. One means of accomplishing this is shown in Fig. 13.5. In this figure the cam follower is about to rise onto the lobe of the cam, which is rotating counter-clockwise. When the follower rises, it will initiate the injection of fuel. For operation astern the reversing cylinder moves the top of the link carrying the cam follower to the right, placing the follower in the correct position, relative to the cam, for rotation in the opposite direction.

Fixed-pitch propellers must be driven in reverse to propel a ship astern. Low-speed engines directly driving fixed-pitch propellers are necessarily direct reversing. Medium-speed engines driving fixed-pitch propellers through reduction gears may be direct reversing, or reversing elements may be incorporated in the gearing. Most high-speed engines driving fixed-pitch propellers use reversing gearing.

CHAPTER 14
DIESEL ENGINE COMPONENTS

This chapter is organized around figures, and is intended to enable features of components to be discussed in greater detail than so far presented.

14.1 Bedplates, bearings, and crankcases

Figs. 14.1 and 14.2 show the classic construction used in large, low-speed engines and in some medium-speed engines, with the **bedplate** at the bottom, carrying **main bearings** for the crankshaft, and a **mounting flange** for bolting the engine to its foundation. The middle section is the **crankcase**, also called the housing or frame. The upper part is the **cylinder block** in which the **cylinder liners** will be inserted. The bedplate, crankcase, and cylinder block are assembled as shown in Fig. 14.2, and are secured by large vertical **tie bolts** that pass through the holes visible in Fig. 14.1. It should be obvious by counting access openings in the crankcase and cylinder blocks that both of these figures are of components for six-cylinder engines. The seventh bay in the bedplate would accommodate the **thrust bearing** in a direct-connected engine. These components may be of steel or cast iron.

Fig. 14.3 is similar to Fig. 14.1, but with more detail. The assembly on the right, which is for three cylinders of a larger engine, shows how components may be constructed in pieces, often to suit available manufacturing capability.

Figs. 14.4 and 14.5 show the bedplate and crankcase of a six-cylinder, low-speed engine with some of the smaller components that will be attached. These include one set of main bearing **shells** (170), **caps** (108, 336), and **studs** (62). The mounting flange and its **hold-down bolts** are accessible through the low row of ports in the sides of the bedplate. The piping pieces in Fig. 14.4 are for lubricating oil.

Fig. 14.6 shows a cast iron bedplate for a medium-speed engine, with main bearing lower shells, and bearing studs, in place.

Fig. 14.7 illustrates the assembly of a main bearing, with a single cap secured by two studs and nuts.

Fig. 14.8 shows cast-iron stationary components for a medium-speed engine that is manufactured in both in-line and V versions. In the in-line version, the crankcase and cylinder blocks are combined in a single casting. The tie rods appear prominently. In both engines, the crankshaft main journals are visible in section, with the main bearing caps bolted down.

Figs. 14.9 and 14.10 also show cast-iron stationary components but for two other designs of medium-speed engines. In these designs the single casting combines bedplate, crankcase, and cylinder jackets. In contrast with the main bearings previously described these main bearings are inverted or **underslung**, with the upper halves cast into the bedplate, and the caps inserted from below. This arrangement is often found in medium- and high-speed engines. Also visible in Fig. 14.9 is the recess in the side of the cylinder block that will accommodate the **camshaft**. In both of the section views at the top of the figure the camshaft **bearing caps** are shown in place.

14.2 Crankshafts

Figs. 14.11 through 14.13 are photographs of **crankshafts**, all of forged steel. Fig. 14.11 shows the bedplate of a nine-cylinder, low-speed engine, with the crankshaft installed and main bearing caps bolted down. Fig. 14.12 is a crankshaft for a four-cylinder, low-speed engine, showing, from the right, the **drive flange** for connection to the propeller shaft, a flange for connection of the **flywheel** and **timing gear**, and the **thrust collar**. Fig. 14.13 shows crankshafts of three different medium-speed engines, all of forged steel. The shaft at the top of Fig. 14.13 has not yet been fitted with **counterweights**. The holes in the journals are lubricating oil passages.

14.3 Crossheads and connecting rods

Contrast the short **crankpin journals** in Figs. 14.11 and 14.12, and those of the middle crankshaft in Fig. 14.13, with the longer crankpins of the other crankshafts in Fig.14.13. The longer crankpins are necessary in most V-engines to accommodate the **connecting rod bottom-end bearings** in pairs, while in-line engines have only a single connecting rod on each crank. Fig. 14.14, a cut-away view of a V-engine, shows the usual arrangement of paired connecting rods on each crank. (The cylinders of each bank are therefore not in the same transverse plane, but are offset axially.) Fig. 14.15 is a section through a medium-speed, V-engine which illustrates an alternative method of joining the power output of the two cylinder banks: the right connecting rod is linked to the big end of the left rod in an **articulated** arrangement. Also prominent in this view are the oil passages in the crankpin.

Figs. 14.16 through 14.20 illustrate **crosshead** assemblies of several low-speed engines. In Fig. 14.16, on the left, is a complete assembly of piston, piston rod, crosshead with **shoes**, and connecting rod, while the right view shows only the crosshead with its shoes, exposing the **crosshead pins** to view. Fig. 14.17 shows this crosshead in its **guides** inside the crankcase of an engine. In this view, the closer guide on the right is cut away to expose the shoe. Note how the guides confine the shoes, and therefore the crosshead and piston, to a vertical motion. In Fig. 14.18 the crosshead is shown sectioned. Additional crossheads and connecting rods are shown in Figs. 14.19 and 14.20. These are similar to each other in attaching the piston rod to the top of the crosshead with a flange, very evident in Fig. 14.19 (rather than by a nut installed from the bottom as in Figs. 14.18 and 14.21) enabling a full-width bearing at the bottom of the crosshead.

14.4 Pistons

Fig. 14.21 shows the **piston** and **piston rod** of a low-speed engine. Two tubes are fitted (only one appears in the drawing, since the other is directly behind) to carry cooling water to and from the inside of the piston. These tubes telescope into stationary pipes in the crankcase as the piston reciprocates (see Fig. 11.12, where the **telescoping tubes** are visible on the right side of the piston). Note how the **crown** in Fig. 14.21 is bored from beneath to allow the cooling water to penetrate close to the hot top of the piston. When the pistons are oil-cooled, the oil is usually carried through internal borings in the piston rod.

Figs. 14.22 and 14.23 are illustrations of **trunk pistons** from three different engines. Note the passages for oil cooling and lubrication. The connecting rods, wrist pins and piston crowns are generally of forged steel. The lower part of a two-piece piston, the **skirt**, is most often of cast iron.

14.5 Cylinders and cylinder heads

Figs. 14.24 to 14.26 show **cylinder liners**. The row of ports in the liner of Fig. 14.25 identifies this as a liner for a two-stroke, uniflow-scavenged engine. In Fig. 14.26, the dark circles at the top of the liner are bored cooling-water passages. The cooling water enters the space between the liner and the jacket at the bottom, flows up on the outside of the liner (which is therefore called a **wet liner**), then enters the bored passages, then flows up to the cooling passages in the head, and exits at the top. Cylinder liners are almost always made of cast iron.

Figs. 14.27 through 14.31 show **cylinder heads** of four-stroke engines. In Fig. 14.27, from the left, are the exhaust valve, the fuel injector, and the air valve. The exhaust valve is shown open, with its spring compressed (compare its length to that of the air valve spring). Note that the exhaust valve seat is cooled by nearby cooling water passages, unlike the air valve seat. Also note how the portion of the head immediately surrounding each valve is a separate entity from the head (identifiable as such by the different crosshatching) and can be lifted out of the head from above, as shown in Fig. 14.28. The removable assembly for each valve is called a **valve cage**, and enables a new valve, complete with its spring, guide, and seat, to be inserted without having to remove the cylinder head from the engine. Contrast this arrangement with the head of Fig. 14.29: although the components are replaceable, in this case the cylinder head must be removed from the engine to replace the valve and its seats.

In Fig. 14.29 only the exhaust valves appear, both operated simultaneously by the single rocker arm and pushrod. Figs. 14.30 and 14.31 show a head from a different engine, with an alternative arrangement for the simultaneous operation of paired valves: in this case the exhaust valves are operated by a Y-shaped rocker arm. Cylinder heads are most often of cast iron, although steel castings and forgings are used in some engines.

14.6 Injection pumps and injectors

Fuel injection pumps and **fuel injectors** are shown in Figs. 11.7, 14.32, 14.33, and 14.34. These pumps are all of the cam-operated, helix-controlled type, widely used with medium-to-large diesel engines. Their operation is described below:

> With reference to either Fig. 11.7 or 14.34, rotation of the **cam** forces the **cam follower** and **pump plunger** into a vertical reciprocating motion. The pump plunger is simply a small piston, and the **barrel**, its cylinder. When the pump plunger is at the bottom of its stroke, with the follower still on the base circle of the cam, fuel flows through the fuel ports to fill the barrel. When the pump plunger moves up, it blocks the fuel ports, then raises the pressure of the fuel trapped in the barrel, until the pressure is high enough to force open the discharge valve, which is otherwise held closed by a spring. As the plunger continues to rise, fuel enters the pipe and displaces fuel into the injector, also at high pressure.
>
> When the requisite amount of fuel has been injected the fuel ports in the pump barrel are re-opened by the recess in the plunger (best seen in Figs. 14.34 and 14.35), causing an almost immediate drop in the pressure of the fuel oil, even as the pump plunger continues to rise. The sudden drop in pressure causes the delivery valve and the injector to close sharply, cutting off the injection.
>
> The requisite amount of fuel oil injected into the cylinder in each cycle is set by the **fuel rack** of the pump. The fuel rack is in turn positioned in response to the power demanded from the engine. To develop more power, more fuel must be injected in each cycle, and therefore, the recess in the pump plunger must re-open the fuel ports later in the stroke of the pump plunger. How this is achieved is shown in Fig. 14.35: the top of the recess in the pump plunger has a helical shape, so that rotating the plunger in its barrel (by moving the fuel rack, which has teeth in mesh with a gear that turns the plunger: see Fig. 14.34) adjusts the point in the plunger stroke at which the fuel ports re-open.

CHAPTER 15
DIESEL PLANT ARRANGEMENTS AND OPERATIONS

15.1 A motorship engine room

Fig. 15.1 shows the arrangement of machinery in the engine room of a bulk carrier. The capsule descriptions that follow should be taken one-by-one, with named equipment located in each of the drawings in which it appears.

Refer to the elevation. The **main engine**, a five-cylinder, low-speed engine, extends from its foundation on the **tank top** through the **floor plates** and the two **flats**. The **engine opening** in these flats is generously clear of the engine dimensions, and is carried up through the flats to provide access to the engine for the traveling **gantry crane**, hung from rails attached to the main deck structure. The principal ladders are run within the engine opening, connecting catwalks around the engine to the flats. Enter the engine room from the accommodation at the ship's main deck on the port side and proceed down the ladder to the upper flat (or take the elevator down). From a vantage point in front of the control room door (refer now to the plan at upper flat) count the five **cylinder heads** of the main engine. Note from the crane rails overhead that the gantry crane can access the **main engine spares** in their storage positions forward on the upper flat, including the spare piston and piston rod that hang within the engine opening. Walk aft to the **landing area** and look up at the hatch in the main deck: both the gantry crane in the engine room and the **stores crane** running athwartships above deck (between the accommodation block and the **casing** in the elevation) can plumb landing areas on each flat and on the floor plates, through the hatches, a feature necessary in handling spares and stores. Also note in the plan view the overhead monorails that can be used to move parts from the upper deck landing area to the workshop and storeroom. Across the engine opening, note that the large **main engine exhaust** duct leads aft and then up, clear of the crane rails and the **oil-fired boiler**, to the **waste-heat boiler** overhead in the casing. At the aft end of the upper flat a door leads aft to the steering gear, and just to starboard of that door is an entrance to the fire-tight **escape trunk** leading from the floor plates up to the main deck, with similar entrances at each level.

Walk around the main engine and forward on the starboard side of the upper flat. Note the compression units for **air conditioning** and **refrigeration**, the **freshwater compression tank** and pumps, and the **hot water heater**, all located just below the accommodation block for which these services are provided. Note the many **fuel tanks** and **lubricating-oil tanks**, most of which are on this flat.

Walk back to the port side, forward of the engine, and continue aft on the port side and enter the **control room**. The **main console** is inboard, and contains controls for the main engine and auxiliary machinery, and for the fuel transfer and ballast systems. Outboard are the electric **switchboard** and distribution panels. Figs. 15.2 and 15.3 are typical views of a console and switchboard.

Descend to the middle flat and go aft to the generator room. Note the three **ship's-service diesel generators** (**SSDGs**), close to each other and to the switchboard and distribution panels above (in the control room). Take note that a **lifting beam** is fitted over each SSDG to assist in engine maintenance and overhaul. Forward of the generator room on the starboard side is the **purifier room**, with the **fuel purifiers**, heaters, pumps, and filters which serve the main engine and generator engines, in close proximity to the heavy fuel tanks above and the diesel oil tanks immediately aft. For convenience, the **lubricating-oil purifiers** are also located here. Note the monorail over the purifiers to assist in maintenance. On the port side of the middle flat starting from aft in the generator room, note the large **starting-air receivers**, and forward of the bulkhead, the **starting-air compressors**, **ship's-service air compressor** and **receiver**, and the **control-air compressor**, **receiver**, and **drier**. Further forward are the **boiler feed tanks** and pumps, and the **sewage treatment plant**.

Continue down to the floor plates. The floor plates are loosely laid on a grid of steel angles several feet above the tank top to leave a **bilge** underneath, for running piping. Since the floor plates are readily removed for access, they continue right up to the main engine. The **propeller shafting** runs aft, at about the height of the floor plates (refer to the elevation again) supported by the **steady bearing**. The **main thrust bearing** is built into the aft end of the main engine. Note the **spare tail shaft**, in brackets on the port side web frames.

Many of the pumps on the floor plate level have to be there to suit their suction requirements, as with the **bilge pumps**, the **fuel transfer pumps**, and the **main-engine lubricating-oil pumps** (which must draw from the **main-engine sump**: refer to the elevation again), and the pumps that require seawater suction, including **seawater circulating pumps**, **ballast pumps**, and **fire pumps**. The **main-engine lubricating-oil coolers** and **jacket water coolers**, and the **central freshwater coolers**, are located at this level to be near the seawater pumps that supply them. The **evaporator** is located nearby because it is heated by main-engine cooling water.

15.2 Preparation and starting

The procedures described here are general, intended to provide background information only, and are not intended to be comprehensive or all-inclusive. Recommended procedures for specific equipment should be sought out and followed.

Prior to starting, an engine that has been shut down for an extended period or that has been overhauled must be thoroughly inspected. Tools, eyebolts and other lifting gear, rags and debris must be cleared from the crankcase, from the air and exhaust manifolds, from the air intakes, and from the general vicinity of the engine.

Fluid levels and condition must be checked. If there is any doubt about the condition of the coolant, the LO, or the fuel, the system should be drained and refilled. On main engines the LO is circulated through the purifier and brought close to its normal temperature.

Where motor-driven cooling pumps are fitted, one of these is started in each circuit and on large engines the cooling water is heated toward the operating temperature. High points and other air traps in the cooling system are vented.

A seawater cooling pump is started and coolers are vented on their seawater sides.

Where independent LO-circulating pumps are fitted, one is started, and a flow to all bearings and service points is assured. On engines with pre-lubrication pumps, these are operated until all bearings receive oil. On crosshead engines, cylinder lubricators are operated manually until all cylinders receive oil.

Drains in air and exhaust manifolds are opened, as are indicator cocks on each cylinder. Where a compression-release device is fitted, it should be used. If the engine is directly connected to the propeller, permission to rotate the engine must be obtained from the bridge, to ensure that lines are clear of the propeller. After it has been confirmed that all hands are clear of the engine and shaft line, the engine is rotated through several revolutions with the turning gear or a barring device to ensure freedom from interference, to help establish oil films at bearings, and to ensure that cylinders and manifolds are free of water. The turning gear or barring device must then be disengaged.

If access permits, turbochargers should be rotated by hand to ensure freedom of rotation.

Fuel systems are primed and vented. While most engines that normally run on HFO are started on distillate fuel and shifted over to HFO only after starting, where the engine will be directly started on HFO, a booster pump is started with the recirculating line open. Steam is lined up to heaters, to steam-traced fuel lines and, where appropriate, to the injector-cooling circuit. The fuel is recirculated through the heaters until hot fuel is present at the injectors, or as close to them as the system permits.

On air-started engines the starting-air receivers are charged and drained.

An inspection is made to ensure that all gage valves are open, that all sensors are in place and connected, and that standby pumps not in use are lined up and operable. Safety devices, including interlocks and alarms, are tested.

Before starting a propulsion engine that is directly connected to the propeller, permission must again be obtained from the bridge.

On reversing engines, the camshaft and starting-air distributor are set in the desired direction. The fuel rack is set to a starting position, usually about 25% of maximum. Starting air is lined up to the starting valve. When the starting valve is opened, the engine will crank through one or two revolutions before firing. As the engine accelerates, the fuel rack is repositioned to the idle-speed setting. However, if the engine is driving a fixed-pitch propeller, the engine will have to be stopped immediately after starting to avoid putting way on the ship, straining mooring lines and bollards. If a decompression device is fitted and used, it is returned to its normal setting as the engine comes to idle speed.

Where possible, the engine should be allowed to warm up at a light load. Where a higher load must be applied immediately, the lowest initial load should be applied and increases should be gradual. The engine manufacturer's suggestions should be followed. Immediately after starting, all systems are checked for proper operation.

On engines that are fitted with attached pumps, but that are started using independent motor-driven pumps, the motor-driven pumps can be stopped when appropriate. If the motor-driven pumps are fitted for automatic standby they must be left correctly lined up for this function.

15.3 Maneuvering

Propulsion engines are usually controlled from the bridge, using direct-mechanical linkage or by pneumatic, hydraulic or electric servomotors, with alternative control stations at the side of the engine and in an engine control room. Once the engine has been prepared for starting control can be turned over to the bridge.

Speed and load control of a diesel engine is achieved by positioning the fuel racks of the injection pumps, which are mechanically linked to operate simultaneously. In some installations, the racks may be positioned directly by the operator, but in most cases the racks are positioned by a governor, whose set point is adjusted by the operator.

Engines that directly drive fixed-pitch propellers are necessarily reversing engines. To meet a stop order the engine itself must be stopped by moving the fuel racks to a zero position. Then, when ordered to run ahead or astern, the camshaft and starting-air distributor are positioned in the appropriate direction, the fuel racks are set to the starting position, and air is admitted to restart the engine. If there is way on the ship when the fuel is cut off in response to a stop order, the engine will continue to be rotated by the propeller until starting air is applied in the reverse direction, or the **shaft brake**, where fitted, is applied.

In fixed-pitch propeller installations where reversing engines are clutched to the propeller shaft (usually at gearing), the engines can be disengaged and warmed up in advance of the maneuvering period. Normal maneuvering may then be as described above, with the clutches engaged throughout, or alternatively, the shaft may be stopped by declutching the engines and allowing them to idle, while a shaft brake holds the propeller shaft stationary. To reverse propeller rotation, however, the engines must be stopped, the camshafts and starting-air distributors shifted, the engines started again in the reverse direction, and the clutches engaged.

Where non-reversing engines drive a fixed-pitch propeller through reversing gearing, a shaft brake must be fitted. Reversing is achieved by reducing engine speed to idle, declutching the ahead gear train, applying the shaft brake to stop the shaft, engaging the clutch on the astern gear train, and then raising engine speed to the desired setting. In this arrangement the clutch movements and shaft brake application would normally be programmed to operate in sequence from a single control lever.

After each use of starting air, the compressors are started, usually automatically, to recharge the receivers. During intensive maneuvering periods the compressors may be running continuously. In most cases the receivers and moisture separators will be fitted with automatic drain traps, but if this is not the case, they must be drained at frequent intervals.

The minimum engine speed is typically 25 to 40 percent of rated rpm and, unless a slip clutch is included in the drive train or electric drive is used, this minimum engine rpm determines minimum shaft speed. To maintain extremely low ship speeds in most ships with fixed-pitch propellers requires that the engines be stopped and started repeatedly or, where clutches are fitted, that they be clutched in and out repeatedly.

If there is a critical-speed range, it is passed through quickly.

Ships with controllable-pitch propellers may be maneuvered entirely by pitch control, usually with the engine speed left constant throughout the maneuvering period, at about 75 to 90 percent of rated rpm.

At the end of a departure maneuvering period, engines are brought to the required power gradually, in increments extended over at least the first hour at sea.

During maneuvering periods, support systems and engine instrumentation are monitored. Depending on the extent and condition of automation present, it may be necessary to make adjustments to keep temperatures and pressures within bounds. Boost blowers on two-stroke engines should be operating at low engine output.

If an engine is idled or run at low loads for an extended period, the charge-air cooling water is regulated (automatically or manually) to maintain the charge air temperature above the intake temperature to avoid condensation in the manifold. Where charge-air coolers are cooled by jacket water, it is frequently possible under these conditions to raise the charge-air temperature well above ambient, to help to ensure prompt ignition and complete combustion, and to avoid cold-end corrosion in the cylinders. For engines normally maneuvered on HFO it may be advisable, during a long idle period, to switch to distillate fuel, but this must be done gradually, usually through a mixing tank, since a sudden reduction in fuel temperature can cause injection-pump plungers to bind.

15.4 Routine operation

During steady running, engine indicators and support systems are monitored regularly, and adjustments are made as required. Even in plants with extensive automation engineers should regularly walk around the engines, listening for unusual noises, looking for leaks and sources of unusual vibration, and checking sight glasses and fluid levels.

Every day at sea fuel is transferred from storage tanks to the settling tank, and where two settling tanks are provided, purifier suction is changed over. Where HFO is used, tank-heating and line-tracing steam is supplied to the storage tank and lines that will be used in the following day's fuel transfer. Water and sediment are drained from settling tanks daily, and other tanks are checked for accumulations at least weekly.

In crosshead-engine installations the cylinder-oil measuring tank is refilled daily, noting the levels before and after refilling in order to calculate the consumption. The cylinder-oil injectors are then adjusted as required.

Moisture is drained daily from charge-air coolers and from the air manifold.

On engines burning HFO the turbocharger turbines are water washed, often every day, in accordance with manufacturer's instructions. The turbocharger compressors will require cleaning less frequently.

At least weekly, or once per passage on vessels on shorter voyages, **indicator cards** (plots of pressure vs. volume, as in Fig. 11.4) are taken on low-speed engines, and maximum-pressure readings on higher-speed engines, in order to confirm that the load is balanced among the cylinders.

Fuel-oil consumption is calculated daily, and lubricating-oil sump levels are checked at least as often.

Self-cleaning purifiers will have to be opened for manual cleaning at intervals determined by experience. Until that experience is gained it may be reasonable to open HFO purifiers weekly, and LO and DO purifiers monthly. Older purifiers that do not eject accumulated sludge will require more frequent cleaning, at least daily for the HFO purifiers and perhaps weekly for the others.

15.5 Stopping

Before stopping an engine normally run on HFO but started on DO it will be necessary to change over to DO, although up to half an hour of operation at high output may be needed for the DO to fill the system through to the injectors. Even for engines that are normally started on HFO, this step may be advisable, as it will simplify any maintenance that will require opening of the fuel system.

When steam is no longer needed for fuel heaters and for fuel-line tracing, it is secured.

On engines with independent cooling-water and LO-circulating pumps, these pumps are kept in operation for about a half hour after the engine has been stopped.

Main propulsion engines cannot be secured until the bridge has given and confirmed the order: **finished with engines**.

Once an engine is secured, fuel pumps are stopped, the starting-air stop valve is shut, and the turning gear is engaged to prevent accidental rotation of the engine. Indicator cocks at cylinders and drain valves at air coolers, at air manifolds, and at exhaust manifolds are opened and left open.

LO purifiers on propulsion engines are kept in operation for 12 to 24 hours after the engine is stopped.

Crankcase doors must not be opened until after the crankcase has cooled down because of the danger of a crankcase explosion.

15.6 Emergency Operation

It is sometimes necessary to put an engine into temporary service with one or more cylinders cut out, or with a turbocharger secured. Detailed procedures and precautions differ from one engine to another, and instruction manuals must be consulted. Only reduced power will be available, and further limitations may exist because of vibrations.

15.7 Maintenance

Diesel engines contain many moving and wearing parts, and concentrated maintenance is necessary to keep them running, and to reduce the likelihood of unexpected failure. Maintenance involves disassembly for inspection and renewal, as well as treatment or replacement of lubricating oil and cooling water. Under the best of circumstances, two approaches to maintenance are usually used together: preventive maintenance, and performance-data analysis, also called condition monitoring or trend analysis.

Preventive maintenance is undertaken according to a schedule, usually set forth by the engine manufacturer. The schedule might list tasks to be undertaken after a certain number of running hours have accumulated. Injectors of engines run on heavy fuel might be pulled for testing and inspection at intervals of 1000 hours. For merchant ship propulsion engines, each cylinder head might be lifted, the piston pulled, the liner inspected, wear measured, a reconditioned piston inserted, and a reconditioned head installed, at intervals of 10,000-to-20,000 hours. For high-performance engines, more frequent maintenance will be recommended.

Performance-data analysis requires that trends in key operating parameters be tracked over time, until the parameter approaches recommended limits, or begins to show a sharp change, thereby providing sufficient warning of imminent failure to allow maintenance to be scheduled. Parameters that may be tracked might include the following:

cylinder exhaust-gas temperatures
turbocharger rpm
boost pressure
indicator cards in low-speed engines, or maximum firing pressures in higher-speed engines
compression pressures
pressure drops across coolers and filters
lubricating oil properties (determined by taking an oil sample for analysis)
specific fuel consumption
exhaust smoke

Advanced measurement techniques are becoming practical and are being increasingly applied in trend analysis. For example, fast-reacting electronic pressure sensors can be fitted to cylinders, enabling analysis of cylinder performance on demand. Some engines are fitted with embedded detectors to measure piston ring and cylinder wear. As more sophisticated measuring techniques and instruments become available and are proven, the number of parameters monitored will increase and the effectiveness of condition monitoring will grow.

In general, main-engine maintenance aboard oceangoing ships is undertaken on a continuous basis, with some work done during each port visit. For large engines this practice usually requires overhauling one cylinder at a time, on a schedule under which the whole engine is maintained over the course of a year or two, after which the cycle repeats. For this purpose, as well as for emergency repairs, major spares are stored on board, ready for use. For example, a fully outfitted cylinder head will be available for installation in place of one in service. A used component withdrawn from the engine is either reconditioned to become the next spare or replaced. An auxiliary engine may be taken out of service for a complete overhaul while the ship is under way, often by a group of technicians traveling with the ship for that specific purpose. A complete set of spare parts would be ordered in advance, while major parts removed from the engine, such as cylinder liners, pistons, and connecting rods, may be returned to the manufacturer for reconditioning. Operators of ships in seasonal trades usually attempt to restrict all planned maintenance to the lay-up period, when complete overhauls might be undertaken.

The premier maintenance tool in a diesel plant is an overhead crane. While a simple overhead beam fitted with a trolley for a chain hoist may suffice for smaller engines, a gantry crane, electrically or pneumatically powered, with mobile remote control, capable of longitudinal and transverse positioning, is normal in larger plants. Maintenance tasks are simplified by the use of special tools and access gear, usually supplied by the engine builder.

CHAPTER 16
PROPULSION POWER TRANSMISSION

16.1 Reasons for transmissions

Propellers are low-rpm devices, with the most efficient propellers designed for speeds below 100 rpm, while acceptable efficiency can still be achieved at up to about 300 rpm. Of the marine prime movers currently in use, only the low-speed diesel can match this speed range, and is therefore directly connected to the propeller. Turbines, which are most efficient at several thousand rpm, medium-speed diesels, which run at speeds of about 400 to 1200 rpm, and high-speed diesels, running at over 1000 rpm, all require speed-reducing transmissions. The primary reason for a transmission, therefore, is to reconcile the high rpm of a steam or gas turbine, or of a medium- or high-speed diesel engine, with the lower rpm of an efficient propeller. The most common type of transmission is mechanical **reduction gearing**. **Electric drive**, in which the prime mover drives a generator whose output then drives a motor on the propeller shaft, is useful in some applications.

A second reason for fitting a transmission is to provide a means of combining the power output of several prime movers to drive a single propeller. Common applications of gearing for this purpose, combined of course with speed reduction, include cross-compound steam turbines, paired gas turbines, and paired medium- and high-speed diesels. Propulsion plants with multiple diesel engines or gas turbines are often electric drive.

A third function that can be provided in a transmission is reversing capability. This capability is sometimes used with medium-speed diesels and gas turbines, and is common with high-speed diesels, when any of these prime movers are geared to fixed-pitch propellers. With electric drive, reversing capability is inherent.

16.2 Principle of gearing

A small gear, called a **pinion**, meshes with a larger gear. To turn the gear through a whole revolution, the pinion must turn a number of times, in proportion to the ratio of the number of teeth on the gear to the number of teeth on the pinion. Since the tooth size at any mesh is the same on both pinion and gear, the number of teeth is proportional to the diameter of the pinion or gear. Therefore, the rpm **ratio** R, of pinion to gear is equal to the ratio of diameter D, of gear to pinion:

$$R = \frac{rpm_p}{rpm_g} = \frac{D_g}{D_p}$$

An example appears in Fig. 16.1, a **single-input, single-output gear set**. The diameter of the gear is about three times that of the pinion, so that a 450 rpm engine connected to the pinion will drive a propeller shaft connected to the gear at 150 rpm.

An item of note in Fig.16.1 is that the teeth are not parallel to the axes of the pinion or gear, but instead align with a helix: **helical gear teeth** continuously engage and disengage from the mesh as the gears turn, yielding steady torque transmission.

A second example is shown in Fig. 16.2(a), in which the faces of the gears are divided by a circumferential gap, into two opposing helices. This **double-helical** arrangement balances the axial thrust that is present in single-helical gearing.

16.3 Multiple-input gearing

Fig. 16.2(b) shows how the power from two input pinions can be combined to drive a single gear. This is the general arrangement of gearing used with paired, medium- and high-speed diesel engines.

16.4 Single-reduction and double-reduction gearing

The gear sets of Figs. 16.1, 16.2(a), and 16.2(b) are **single reduction**: between each input and the output, there is only a single mesh. The maximum rpm ratio with single-reduction gearing is about 20:1, although difficulties are encountered above about 12:1. To achieve the high ratios necessary for turbines, **double-reduction** gearing is used. An example of a **double-input**, double-reduction gear set is Fig. 16.2(c): each input, or high-speed, pinion is in mesh with a first-reduction, or high-speed, gear, which is coupled at its shaft to the hub of a second-reduction, or low-speed, pinion. The second- reduction pinions are in mesh with the second-reduction gear, called the **bull gear**. In this example, the flange on the bull gear shaft, to the left of the gear, is bolted to the propeller shafting; the thrust collar is forward of the bull gear.

Another example of double-input, double-reduction gearing is shown in Fig. 16.3, with a cross-compound turbine set. The overall gear ratio is the product of the first and second reduction ratios. Tracing the path from the high-pressure turbine, if the first reduction ratio is 8:1, and the second reduction ratio is also 8:1, the overall ratio is 64:1, so that a high-pressure turbine running at 6400 rpm will drive the propeller shaft at 100 rpm. The low-pressure turbine path shows a larger high-speed pinion and a smaller high-speed gear, resulting in a lower first reduction ratio for the low-pressure turbine, commensurate with its lower rpm. If the low-pressure turbine first reduction ratio is 5:1, the overall ratio for the low-pressure turbine is 40:1; with the propeller shaft at 100 rpm, the low-pressure turbine will be running at 4000 rpm.

Double-input, double-reduction gearing of the type described in these examples, usually called **articulated reduction gearing**, is common for cross-compound steam turbines up to about 20,000 shp.

16.5 Multiple torque paths

Fig. 16.2(d) shows a **locked-train**, or **dual torque-path**, double-reduction, double-input gear set. What distinguishes this set from the articulated gear sets examined above is that each input pinion is simultaneously in mesh with two high-speed gears. The input power (and torque) at each input pinion is thus divided, and is delivered to the bull gear in two paths. Since each path subjects the teeth to only half the force they would otherwise carry, high power can be transmitted through smaller elements, resulting in a gear set that can be smaller, lighter, and usually less expensive than an articulated set would have to be to transmit the same power. Locked-train, double-reduction gearing is common for steam turbines of 20,000 shp or more, and for gas turbines.

16.6 Quill shafts

A **quill shaft** is a shaft which passes through the center of a pinion or gear, as shown in Fig. 16.4, to be attached to the pinion or gear only at the far end, usually through a **flexible coupling**. The arrangement gives the torsional flexibility of the length of the shaft, yet permits the coupled elements to be close together. Quill shafts are normally incorporated in all double-reduction gearing, coupling each high-speed gear to its low-speed pinion.

16.7 Clutches

Clutches are used to connect or disconnect prime movers from their loads, usually while one or both are rotating. Clutches are not normally fitted with steam turbines or low-speed diesels, but are usually provided with geared diesels and gas turbines.

Many different types of mechanical, hydraulic, pneumatic, and electric clutches are in use. Pneumatic clutches of the type shown in Fig. 16.5 are very commonly used with medium- and high-speed diesels. The inner drum is driven by the engine, while the outer drum is connected to the input pinion of the reduction gear. The outer drum carries an inflatable, tire-like tube, called a gland, which is piped to a compressed-air supply via the rotating coupling at the center. The inside circumference of the gland is covered with friction shoes. When the tube is inflated, the shoes grip the inner drum, forcing it to rotate. Once engaged, the drums and gland rotate as a unit, without slipping.

16.8 A gear set for diesel engines

A single-reduction, double-input gear set for paired, medium-speed diesel engines is shown in Fig. 16.6. Each engine is coupled to a quill shaft, which passes through the pinion to the clutch at the far end. The main thrust bearing is forward of the bull gear. Since no reversing elements are incorporated in this gear set, then either the engines are direct reversing, or the propeller is controllable-pitch.

16.9 Gears vs. electric drive

Gearing is very efficient, with about 98% to 99% of the input power transmitted to the output shaft (the rest is lost to friction), while the highest efficiency of electric drives barely exceeds 90%. See Fig. 24.22. Gearing is also lighter and less expensive. However, electric drive offers certain advantages:

- precise speed control
- rapid reversing
- high torque at low rpm
- ability to combine the output of multiple prime movers
- flexible machinery arrangements (prime movers can be located far from the propeller shafts)
- flexible load management (useful where the mission or trade power requirement rivals the propulsion load)

Where these advantages are important, electric drive may be preferred to gearing. Examples include research ships, icebreakers, cable layers, ferries, and passenger ships.

CHAPTER 17
SHAFTING, BEARINGS, AND PROPELLERS

17.1 Shafting arrangements

Fig. 17.1 shows a typical arrangement of shafting for a single-screw merchant ship with geared machinery, with the engine room separated from the after peak bulkhead by one cargo hold and a shaft alley. Proceeding aft from the bull gear, bearings numbered 1 and 2 are journal bearings built into the gear case to support the bull gear. The thrust shaft and its integral collar are aft of the gear case in this example, but could equally well be forward, in which case the thrust shaft and collar would be integral with the bull gear shaft, as in Fig. 17.3. Each section of shafting is flanged and bolted to the next. The first section of **line shafting** penetrates the after engine room bulkhead and enters the shaft alley. Because the bulkhead is water tight, the shaft penetration is sealed by a packing gland. Each section of line shafting is supported by a **line-shaft, or steady, bearing**. The last section of shafting in this example is the **tail shaft** or **propeller shaft**, which penetrates the hull and carries the propeller. The tail shaft and propeller are supported by bearings in the **stern tube**.

The sections of shafting are forged steel and are generally solid. Hollow shafting is used for controllable-pitch propellers or to reduce weight when shaft diameters are increased to make them stiffer.

Fig. 17.2 shows the shafting arrangement for a ship with a low-speed diesel, in an engine room that is fully aft. There is no shaft alley. Because of the short shaft line, there is only a single section of line shafting and a single line-shaft bearing. The thrust bearing is built into the aft end of the engine, with the thrust collar integral with the crankshaft.

The arrangement shown in Fig. 17.3 is often used for high-speed, single-screw ships, and for ships with multiple propellers, where the shaft penetrates the hull well forward of the propeller. The stern-tube shaft is therefore separate from the propeller shaft. The propeller shaft and propeller are supported by the **strut** bearing.

Fairwaters are casings of light construction, used for streamlining. They may be of steel plate or fiberglass, and may be secured to the stern tube, the strut bearing, the shafting, or, in the case of the cap over the propeller nut, to the hub of the propeller. Rotating fairwaters are often filled with wax to exclude water. The fairwater fitted to close the gap just forward of the propeller, between the propeller hub and the aft end of the stern tube or strut bearing, is the **rope guard**.

17.2 Line-shaft bearings

Line-shaft bearings, also called steady bearings and plummer blocks, support the weight of the shafting. An example is shown in Fig. 17.4. These are usually self-contained, with an oil sump in the bottom of the housing. Because of the low rpm, they can be lubricated by **oil rings**, which are larger than the shaft diameter, and loose, so that they hang into the oil sump. As the shaft and ring rotate, the ring carries oil to the top of the shaft, where it spills out to lubricate the journal and bearing. In some designs an oil disk, clamped to the shaft, is used instead of the oil rings. Normally, the bearing will be cooled adequately by convection from its surface, but an emergency water-cooling line is usually provided.

The bearing housing may be of cast iron or steel. The bottom of the bearing housing is bolted to a pedestal, supported from the ship's bottom structure. The housing is split horizontally to facilitate disassembly. The shells are usually steel, lined with babbitt.

The spherical seat of the bearing in the figure enables it to adapt to the shaft alignment as the hull flexes in different loading conditions, but non-aligning bearings are also common. In some applications there is only a bottom half-bearing within the housing, enabling the shaft line to deflect upwards. Bearings so designed are called spring bearings.

17.3 Stern tubes

The stern tube is a steel tube extending inboard from the stern post or hull penetration to the closest water-tight bulkhead. Renewable bearing bushings or shells are inserted into the stern tube. Stern tubes are designed for bearings lubricated by either seawater or oil. Oil lubrication is preferred for most merchant ships because oil-lubricated bearings do not normally wear enough to disturb the alignment of the shafting, while water-lubricated bearings wear sufficiently to require replacement after several years of service. Water-lubricated bearings also have more friction, and require that the shaft be protected against corrosion by a bronze liner.

Oil-lubricated stern-tube bearings usually have cylindrical shells of cast iron or bronze, lined internally with babbitt to form the bearing surface. The stern tube is kept full of oil and therefore requires a seal at each end to retain the oil. The after seal also prevents ingress of seawater.

For water-lubrication, the bearings consist of cylindrical bronze bushings, as in Fig. 17.5, which are cast with longitudinal grooves, in which are inserted rubber strips to form the bearing surface. The strips can be of materials other than rubber, including a hard composition called **phenolic resin**. At one time a hard wood called **lignum vitae** was used. At the forward end of the stern tube is a stuffing box and packing gland, which is slacked off slightly when underway, allowing water to fill the stern tube to lubricate the bearings.

When the stern tube supports the weight of the propeller, as in Fig. 17.1, two bearings are generally fitted, with the after bearing more heavily loaded, and therefore longer. Sometimes the forward bearing is eliminated. Stern tubes used with strut bearings, as in Fig. 17.3, usually have only one bearing.

17.4 Strut bearings

Strut bearings may be water or oil-lubricated. Either way, the construction is similar to that of stern-tube bearings.

17.5 Propeller shafts and stern-tube shafts

A propeller shaft must be inserted through the stern-tube or strut bearing, and can therefore have a permanent flange at only one end. Propeller shafts for fixed-pitch propellers are shown in Figs. 17.1 and 17.3. There is a flange at the forward end, and the propeller is fitted to a **taper** at the aft end. This taper is a precise fit to a matching taper bored in the propeller hub. In older construction, the hub is further secured against rotation on the taper by a longitudinal **key**, but most modern propellers are press-fitted onto the taper without a key. In either case, a nut threaded onto the aft end of the shaft keeps the propeller from backing off the taper. The thread of the nut is opposite-hand to the forward rotation of the propeller, and it is further secured by a **keeper**.

The stern-tube shaft for strut bearing arrangements, like the propeller shaft, can have a permanent flange at only one end, since it must be inserted through the stern-tube bearing. In the example of Fig. 17.3, with the flange outboard, the inboard end is coupled to the line shafting by a **muff coupling**. A muff coupling, one example of which is shown in Fig. 17.6, consists of a sleeve overlapping, and secured to, the ends of both shafts. In the figure, as in most modern construction, the sleeve is secured by press fitting to both shafts, but older coupling sleeves are secured by keys.

The steel shafting must be protected from corrosion by seawater. Propeller shafts and stern-tube shafts for water-lubricated bearings are usually fitted with bronze liners. Oil-lubricated shafts are bare through the stern tube, but are fitted with a short, stainless-steel sleeves at the seals. In either case, for propeller shafts, the joint between the forward end of the propeller hub and the aft end of the liner or aft sleeve is sealed by a rubber O-ring. For shafting with strut bearings, exposed surfaces are covered by rubber, plastic, or fiber glass.

17.6 Propellers: fixed-pitch and controllable-pitch

Most ships have fixed-pitch propellers. These propellers are usually cast in one piece, blades and hub together, usually of aluminum bronze, although stainless steel is used in some applications. Compared to controllable-pitch propellers, fixed-pitch propellers are much less expensive, are simpler and therefore more reliable, and are more efficient at their design rpm.

Controllable-pitch propellers have a hydraulically operated mechanism in the hub that rotates the blades, all together, to change the pitch. An example is shown in Fig. 17.7. Changing the pitch enables the thrust to be varied without changing rpm, from full ahead to full astern. Controllable-pitch propellers offer certain advantages:

- maximum thrust availability at low ship speed (useful for tugboats, trawlers with diesel engines, and icebreakers with diesel engines or gas turbines)
- ability to match to different prime movers geared to the same shaft (useful for single-engine operation of ships with multiple diesel engines or gas turbines)
- reversing capability for non-reversing engines (especially useful for gas turbines)
- rapid maneuvering without stopping the engines or using clutches
- constant rpm, independent of ship speed (useful when shaft-driven generators are fitted)

Where these advantages are important controllable-pitch propellers may be justified. Examples include vessels of all types and sizes.

Controllable-pitch propellers are flanged to their propeller shafts. The shafting is hollow to accommodate the hydraulic lines or mechanical link rods that operate the hub, and that connect to a control box located forward on the shaft line, most often at the very forward end.

17.7 Withdrawal of propeller and stern-tube shafts

At intervals of three to five years, propeller and stern-tube shafts must be withdrawn for inspection. This job is almost always done in dry dock. For the typical arrangements of Figs. 17.1 and 17.2, where the propeller shafts are drawn inboard, the fairwater and propeller nut are removed, and the propeller is released from the taper, to be supported in place by staging or by chainfalls hung from padeyes welded to the hull. The last section of line shafting, with its bearing, is removed using overhead lifting gear which is usually provided (see Fig. 15.1), and set aside on cradles and landing areas specially prepared for this purpose. Seals or packing glands are cleared away. The lifting gear, or, in its absence, wood blocking, is then used to support the shaft as it is pulled inboard by a chainfall. If the shaft must be repaired or replaced, it is removed through a hole cut in the side shell, in an area originally designated for this purpose.

For the strut bearing arrangement of Fig. 17.3, the key to the procedure is the removal of the propeller shaft. After the propeller is freed and supported, and the fairwaters are removed, lifting gear is used to take the weight of the propeller shaft off the strut bearing, which is then removed from the strut by sliding it along the shaft. This procedure leaves enough radial clearance in the strut bearing housing for the shaft to be tilted, as shown, and pulled forward and lowered to cradles on the floor of the dry dock. The stern-tube shaft can then be pulled outboard, using lifting gear and chainfalls, after disconnecting the inboard coupling and clearing packing glands or seals.

Controllable-pitch propellers are usually flanged to their shafts, requiring that the propeller shaft be pulled aft. To enable this to be done, the propeller must be moved aside after it is freed from the shaft and supported clear of the work area. Often, the rudder is also in the way, so that it may have to be removed as well. After clearing away fairwaters, glands, and seals, the forward coupling of the propeller shaft can then be disconnected, and the shaft pulled aft, supported by chainfalls or staging.

CHAPTER 18
GAS TURBINE PLANTS

18.1 Characteristics of gas-turbine plants

Gas turbines can be divided between those derived from aircraft jet engines, called **aircraft-derivative gas turbines** (ADGTs), and those originally designed for stationary power production, called **heavy-duty gas turbines** (HDGTs). Although both types run on the same basic thermodynamic cycle, and have the same general configuration, they differ in performance parameters and design philosophy.

The emphasis in aircraft engines is on high power-to-weight ratios and therefore on high performance, often resulting in expensive materials and manufacturing processes, limited component lives, and frequent rebuilds relative to the simpler but more robust heavy-duty designs. In the aircraft application there is no consideration of any fuel but the cleanest distillate, or of add-on equipment to utilize waste heat. These characteristics carry through to the ADGTs used for ship propulsion. However, the high power density, low specific weight, and stand-alone features of ADGTs are very attractive in high-performance naval vessels (and some commercial vessels as well), where the relatively frequent rebuilds and high fuel-quality requirement are acceptable. At present, ADGTs dominate in the propulsion of surface combatants up to the size of cruisers and small aircraft carriers, in most of the world's navies.

HDGTs are designed with less emphasis on low weight, which results in easier operating conditions and more robust, simpler components, with longer lives and a higher tolerance for fuel quality. The lower performance parameters for which HDGTs are designed encourage the use of added-on heat-recovery equipment, but result in high fuel consumption for simple-cycle HDGTs, relative to ADGTs. As can be seen from Table 18.1, the heat-recovery HDGT plant does not have the advantages of the ADGT in power density and specific weight; nor does it have the advantages of the diesel plants in fuel consumption or fuel-quality tolerance. As a result, although a number of successful merchant ships were built with HDGTs, only a few of these remain in service today.

Table 18.1
Gas turbine plants vs. other prime movers

plant type	steam turbine	medium-speed diesel	high-speed diesel	regen-cycle HDGT	ADGT
specific weight, engine only, in kg/kW	NA	7-25	3-7	7-13	0.2-1
power density, engine only, kW/m^3	NA	40-80	80-300	120-400	150-750
specific weight, whole plant, kg/kW	20-90	25-90	13-30	30-90	7-20
power density, whole plant, kW/m^3	1.2-5	1.2-4	4-10	1.2-4	6-15
fuel quality/maintenance (mod = moderate)	low/low	high/mod low/high	high/mod	high/mod mod/high	high/high
specific fuel consumption at rating, g/kW-h	270-330	180-200	200-230	200-270	230-330

18.2 Gas turbines vs. other prime movers

Advantages of gas turbines over other marine prime movers are:

- high power density and low specific weight, especially for ADGTs
- manufacture as standardized units, with broad experience in other applications
- little on-board maintenance

rapid start and assumption of load
low exhaust emissions, especially of NOx

Disadvantages are:

sensitivity to fuel quality
high fuel consumption, especially at less than full power
high air and exhaust flows
sensitivity to losses in air and exhaust ducts
reduced performance at high ambient air temperatures
high maintenance cost, especially for ADGTs
high thermal profile, because exhaust temperatures and flow rates are high (this is a naval consideration, where a high thermal profile leads to easy detection by satellite or heat-seeking missiles)
no inherent or simple reversing capability
output at high rpm, which requires speed-reducing transmissions for propeller drive

18.3 Principles and major components of a gas turbine

Figs. 18.1 and 18.2 illustrate a **two-shaft**, General Electric LM2500, aircraft-derivative gas turbine, a type frequently found in marine propulsion service, delivering up to 24 MW. Air is drawn into the **compressor** at the left, from an intake located well above sea level. The compressor consists of a number of fan-like, axial-flow stages, each similar to a turbine stage run in reverse, with moving blades on the rotor, and vanes fixed in the casing. This compressor has sixteen stages, and runs at about 9,000 rpm; the compressor discharges the air to the combustor at about eighteen atmospheres. The ratio of compressor-discharge pressure to intake pressure is called the **pressure ratio**.

In the **combustor**, fuel is burned continuously, raising the temperature of the gas leaving the combustor, and therefore supplied to the nozzles of the high-pressure turbine, to about 1250°C. In two-shaft machines such as this, the **high-pressure turbine** drives the compressor, but not any external load. This high-pressure turbine has two stages and exhausts at almost six atmospheres and 850°C. The assembly of compressor and high-pressure turbine, with casings and combustor, is called the **gas generator**, since its purpose is to supply this high-energy gas to the low-pressure turbine.

The **low-pressure turbine** is also called the **power turbine** or **load turbine**. In this example, it has six stages and runs at a maximum of 3600 rpm, suitable for driving a propeller via double-reduction gearing or electric drive. The exhaust from the low-pressure turbine is just above atmospheric pressure, at about 550°C.

Each of the two shafts is supported by two bearings, which are carried in housings supported by internal struts from the casing. The intake-end bearing housing of the compressor shaft, and the exhaust-end bearing housing of the power turbine also act as thrust bearings. The unit is supported from a rigid bedplate, as shown in Fig. 18.3, by external struts at the exhaust end, but by links from a gantry in way of the inlet. The external struts hold the turbine in alignment with the connected load, while the gantry enables the casing to expand as it warms up. One of the internal struts supporting the intake-end bearing housing also houses an auxiliary shaft, perpendicular to the compressor shaft and connected to it by bevel gears inside the bearing housing, which drives the accessory package. This package includes lubricating oil and fuel pumps, as well as a starting motor.

The entire unit is enclosed as shown in Fig. 18.4. The overall dimensions are about 2.4m x 2.6m x 8m, and the weight is some 5000 kg. At an output of 24 MW, this unit has a specific output of almost 5 kW/kg, and a power density approaching 500 Kw/m^3.

18.4 Temperature limits, excess air, bleed-air cooling

There is an upper limit to the temperature of the gas at the turbine inlet, and therefore at the combustor exit, since the first-stage nozzle vanes and blades have to be able to withstand this temperature. The value cited for the example above, 1250°C, is a practical limit for turbines with nozzles and blades that are internally cooled. In order to hold the gas temperature down, large quantities of air are used in the combustors, over 300% excess air in the example cited.

Internal cooling requires hollow vanes and blades. The cooling air is bled from the compressor and, for the blades, is led the through internal passages in the rotor. These provisions are among those features that make gas turbines of advanced design more complex to build and maintain.

18.5 Waste-heat recovery vs. simple-cycle gas turbines

The high temperature, and therefore high energy, of the exhaust gas makes waste-heat recovery attractive in the right circumstances. In **simple-cycle gas turbines** the exhaust heat is not recovered, or may be recovered for auxiliary use. If the waste-heat is returned to the cycle, or otherwise added to the shaft power of the gas turbine, it will improve the efficiency and fuel consumption significantly, especially at part-power levels. However, this improvement must be traded-off against the added weight, the reduced power density, and the higher complexity, all of which are in contradiction to the usual reasons for using gas turbines in the first place. In addition, initial cost and maintenance requirements will be increased.

Gas turbines with pressure ratios up to about twelve can use **recuperative cycles** (also called **regenerative cycles**), illustrated in Fig. 18.5(b). The hot exhaust gas is used to heat the air before the air enters the combustor. This cycle cannot be used with high pressure-ratio machines because the temperature of the compressor-discharge air, which increases with the pressure ratio, approaches or exceeds the temperature of the exhaust gas.

Gas turbines with high pressure ratios can use recuperative heat recovery only in combination with **intercooling**, as in Fig. 18.5(c): air is taken from the compressor part way through, cooled in the intercooler (by circulating water), then returned to the compressor. The resulting temperature of the compressor-discharge air is then low enough, despite the high pressure ratio, to absorb heat from exhaust gas in a **regenerator** or **recuperator**.

All gas turbines can be used in COmbined Gas-And-Steam, or **COGAS**, cycles, illustrated in Fig. 18.5(d), in which the exhaust gas is used to generate steam in the waste-heat boiler, with the steam used in a steam turbine. Gas turbine and steam turbine power output can then be combined at reduction gearing or by electric drive.

18.6 Maintenance by replacement

Steam plants and most diesel engines are maintained and repaired in place, using the ship's crew for some tasks performed on a continuous basis with the ship underway, while major jobs require the ship to be out of service for extended intervals. The concept of **maintenance by replacement**, in which a whole unit is removed from the ship and exchanged with an identical but already rebuilt unit, is particularly adaptable to the ADGTs in naval vessels, for these reasons:

- military crew is relieved of many routine maintenance tasks better handled by experienced civilians in specialized facilities ashore
- time out-of-service is minimized
- the large air intake duct provides a natural removal route, as shown in Fig. 18.6
- the large number of identical units in service minimizes the number of spare units required
- an ADGT, whether the gas generator alone, or the complete engine, is compact, light, and easily handled

The ship and gas turbine installation are designed for maintenance by replacement from the beginning, with easily disconnected piping, wiring and mechanical attachments, and rails or padeyes, and hatches in the ducts.

18.7 Compressor and turbine types

The axial-flow compressor and turbine stages are typical of medium and large gas turbines, but small units often use centrifugal compressors and radial turbines.

18.8 Number of shafts

In addition to two-shaft gas turbines, there are **single-shaft** and **three-shaft** machines. Single-shaft machines are common for generator-drive, for both turbo-electric propulsion and for ship's services, with a single, multi-stage turbine driving both the compressor and the generator. Fig. 18.7 shows the arrangement of three-shaft machines: the compressor stages are divided into low-pressure and high-pressure groups, driven, respectively, by the intermediate and high-pressure turbines, with the exhaust from the high-pressure turbine passing to the low-pressure, or power, turbine.

18.9 Combined prime movers

A gas turbine is often combined with another gas turbine or a diesel engine to drive the same propeller shaft. Some of the more common configurations are described below:

- two gas turbines, usually identical, operated together to achieve high ship speeds, in an arrangement called **COGAG**, for COmbined Gas turbine And Gas turbine

- two gas turbines, one a small engine of low rating for cruising speeds, the other a high-powered gas turbine for high speeds, in a **COGOG** arrangement, for COmbined Gas turbine Or Gas Turbine

- a diesel cruise engine, and a high-powered gas turbine for high speeds, in a **CODOG** arrangement, for COmbined Diesel Or Gas turbine

Usually, a controllable-pitch propeller is necessary with these arrangements in order to match the different operating conditions.

CHAPTER 19
NUCLEAR PROPULSION

19.1 Background

There are compelling advantages in nuclear propulsion:

Nuclear fuel provides an enormous amount of energy from a small mass of fuel, enabling even a large, fast ship to have an almost unlimited range of operation.

Nuclear reactions do not require combustion air, enabling true submarines, of large size and high speed, to operate without surfacing, from the beginning to the end of their missions.

There are serious disadvantages to nuclear propulsion:

Radiation is released. Radiation is hazardous to life, can be permanently contaminating, and can alter the characteristics of materials.

Because of the radiation hazard, and because of the concentrated energy, the consequences of a failure are immense. At the same time, the potential for minor problems to escalate to catastrophic levels is high, with much of the most critical equipment inaccessible in operation because of the radiation hazard, and because of the long operating periods made possible by the almost unlimited range. Therefore, equipment must be carefully designed and constructed, using the highest quality materials, with multiple redundant systems, and must then be operated and maintained by highly trained personnel, all under the close scrutiny of regulatory agencies.

Problems of disposal of spent fuel, radioactive waste, and, ultimately, of the radioactive scrap when the ship has reached the end of its life, have yet to be solved.

For these reasons acquisition costs and operating costs of nuclear-powered ships are very high, while the scrap value or, more likely, the disposal cost, is unpredictable.

Nuclear-powered ships are not welcome in many ports.

There are over 200 ships in service today with nuclear propulsion. The vast majority of these (90%) are submarines in the Russian, American, British, French, and Chinese navies. Most of the nuclear-powered surface ships are aircraft carriers (American and French) and Arctic icebreakers (Russian). All of these are steam ships, with plants based on **pressurized-water reactors** (PWRs). Fig. 19.1 is a schematic of a PWR-based plant.

Because of the nature of these applications there is little information available in the open literature on modern marine nuclear plants. On the other hand, there is a great deal of published information on the Nuclear Ship *Savannah*, which is the 22,000 shp experimental merchant ship operated from 1961 to 1971 by the US Atomic Energy Commission. Despite the age of the plant, and the fact that the technology has advanced, the basic concepts have not changed, and frequent reference will be made to this plant.

19.2 Fission, heat release, and the chain reaction

When an atom of nuclear fuel, usually **uranium**, is struck by a **neutron** moving at just the right speed, it will split, or **fission**. Each such fission reaction releases a finite amount of heat, and also releases two or three neutrons. The practical application of nuclear energy requires two goals to be achieved:

The heat released by each fission reaction must be collected for conversion to mechanical power.

At least one of the neutrons released in each fission must go on to cause the fission of another uranium atom, in a self-sustaining process called a **chain reaction**.

19.3 Pressurized-water reactor

A nuclear reactor is arranged to support a continuous chain reaction, and to harness the energy released. The fuel is concentrated in the **core** of the reactor, most often in the form shown in Fig. 19.2, where the fuel is processed into **uranium dioxide**, molded into ceramic **pellets**, each about 3/8" in diameter by 3/4" long. These pellets are sealed in **zircaloy** tubes to form **fuel rods**. Zircaloy is an alloy of zirconium developed for this purpose. The rods are fastened into square **fuel assemblies**, as in Fig. 19.3, typically of a hundred or more rods, and the assemblies, in turn, are fastened together, as in Fig. 19.4, to form the core of the reactor. The core thus appears as a forest of closely-spaced, uranium-filled fuel rods.

In service, the core, which is in the lower half of the **reactor vessel**, as in Fig. 19.5, is immersed in the pressurized water that fills the vessel. The water acts as a **moderator**, slowing down the neutrons released in each fission reaction until they are at just the right speed to cause more fission reactions. In order to control the number of neutrons in circulation, or the **flux**, some of the neutrons are absorbed by the **control rods**, which are covered with neutron-absorbing material, and can be moved in or out of the core to regulate the rate of the reaction.

The heat must be removed to prevent the reactor components from melting. This is done by the water, which, in addition to acting as the moderator, is also the **primary coolant**, filling the **channels** between fuel rods, while circulating upward, along the rods, and absorbing heat from the outer surface of each rod.

The highest temperature in the reactor is at the centerline of the fuel pellets near the middle of the core. The uranium-dioxide pellets melt at about 4500°F and the maximum centerline temperature in normal service must be kept well below this temperature to preclude melting during transients and in accidents. To prevent structural failure the maximum temperature of the zircaloy tubing of the fuel rods must not exceed about 800°F in normal service. If this maximum temperature is reached at the inside surface of the tube, the corresponding outside surface temperature is about 650°F. As a result the primary coolant, as it rises through the channels near the center of the core, is heated to perhaps 600°F. In most of the channels, away from the center of the core, lower temperatures prevail, so that the average temperature of the primary coolant as it leaves the reactor vessel is lower than 550°F.

The primary coolant enters the reactor vessel at about mid-height and is guided by baffles to the bottom of the core. It is already at a temperature of almost 500°F at this point, at which it enters the channels between the fuel rods. For the water to cool the fuel rods adequately, and act as an effective moderator and coolant, it must be kept from boiling. Boiling is prevented by pressurizing the primary coolant to about 1800 psi. The reactor shell is therefore a thick-walled, heavy, high-pressure vessel. In the 22,000 shp *Savannah* plant, the reactor vessel is about 9 ft in diameter, 27 ft high, and 6.5" thick.

The principal means of control of the reactor is the control rods, which are moved in or out of the core by the drive motors above the reactor vessel. When all of the control rods are fully inserted there is a very low level of activity since most neutrons are absorbed by the rods before they can cause fission reactions. In operation the rods are positioned to absorb just enough neutrons to maintain the chain reaction at a level that produces sufficient heat to generate the steam needed to match the demand. If more steam is demanded the rods are further withdrawn so that they absorb fewer neutrons, and the reaction escalates to a higher level. If some rods are driven part way into the core the reaction diminishes. If enough rods are driven all the way in, the chain reaction dies out.

19.4 Primary circuit components

Refer to Figs. 19.1 and 19.6. The primary coolant leaves the reactor vessel at less than 550°F, and proceeds to the **steam generators**. Here, the primary coolant circulates through the insides of the tubes, giving up heat through the tube walls to the **secondary-circuit** water outside of the tubes. The primary coolant finally leaves the steam generators at about 500°F. The water outside of the tubes, supplied by the feed pumps in the secondary circuit, boils as it absorbs the heat from the primary coolant. The steam so generated rises to the upper parts of the steam generators, and exits via the moisture separators and the dry pipes. The maximum secondary-circuit steam temperature is limited by the temperature of the primary coolant. Since the primary coolant enters the steam generators at less than 550°F, it follows that the steam generated cannot be much hotter than perhaps 475-500°F. Because there is little to be gained by superheating steam at these relatively low temperatures, the saturated steam is usually sent directly to the turbines. There are usually two or more steam generators for each reactor.

Usually, there is at least one **primary coolant pump** for every steam generator. The pump circulates primary coolant through the reactor and steam generator. These are usually large, motor-driven, vertical centrifugal pumps.

The **pressurizer**, shown in Fig. 19.7, is a vertical pressure vessel, used to maintain the pressure in the primary circuit. A steam bubble is maintained in the upper part of the pressurizer: a water-spray nozzle in the steam space is used to condense steam, thereby reducing the pressure in the primary circuit; while heating coils below the water level can be used to generate steam, thereby raising the pressure. One pressurizer serves all of the primary circuits for each reactor.

19.5 Containment and shielding

The **containment** is an airtight pressure vessel enclosing the primary-circuit components, including the reactor, the steam generators, primary coolant pumps, pressurizer, and piping. The containment is designed to contain the radiation that would be released in the event of a catastrophic failure of the nuclear system, and to remain intact if the ship sinks. For marine plants, the containment is likely to be a steel shell. A plan view of the *Savannah* containment is shown in Fig. 19.8. The containment is normally sealed while the reactor is operating.

Shielding is used to protect personnel from radiation. The first level of shielding is the reactor vessel itself, and generally, a tank of stationary water in which it is partially immersed, visible in Fig. 19.8. In this example the **shield water tank** is covered with a layer of lead. Additional shielding surrounds the containment vessel, generally layers made up of concrete, lead, or polyethylene.

19.6 Secondary cycle

Steam conditions for the secondary cycle are limited by the temperature of the primary coolant, as indicated above. Saturated steam at the steam-generator outlet at 475 to 500°F will have a pressure of about 500 psi. Apart from the relatively low steam conditions and the additional facilities for moisture separation, most of the secondary-cycle components are similar to components in fossil-fired steam plants. Of course, the condensers and circulating-water pumps and piping for submarines must be designed for the high seawater pressures at submerged depths.

19.7 Arrangements, and comparison with conventional propulsion plants

Figs. 19.9 and 19.10 show the arrangement of the NS *Savannah*. The principal features, a reactor room forward of the engine room, protected by ship's structure from grounding and collision damage, are typical of nuclear ships, including submarines.

Table 19.1 compares the machinery weights of the *Savannah* with those of conventional ships. Some conclusions can be drawn from these data, and from the *Savannah* arrangements:

> When the fuel weight is included nuclear propulsion plants are not necessarily heavier than other propulsion plants.
>
> The containment and shielding account for more than half the weight of the nuclear plant, and much of this weight is concrete. If weight must be limited, using lightweight shielding materials such as polyethylene in place of concrete might reduce the total significantly.
>
> In terms of volume, the reactor and containment occupy prime space within the hull. While a similar volume might be occupied by fuel in an oil-fired ship, fuel-oil storage can be disbursed to less useful spaces.

Table 19.1
Machinery weights of NS *Savannah* vs conventionally powered ships

NS *Savannah*, 22,000 shp:	weight. tons	subtotal, tons	total, tons
reactor, including fuel	130		
steam generators	100		
primary coolant pumps, piping, pressurizer and auxiliaries	45		
primary coolant	35		
total reactor and primary plant		310	
containment and shielding:			
steel	430		
polyethylene	70		
lead	500		
concrete	1050		
shield water	50		
total containment and shielding		2100	
secondary plant, about		1300	
total, including fuel			3700
Oil-fired steam ship of 22,000 shp:			
machinery with contained water and lubricating oil		1500	
fuel for 10,000 miles		2500	
total, including fuel			4000
Motorship, with low-speed diesel engine of 25,000 bhp (to maintain 22,000 shp in service):			
main engine		800	
remaining machinery with contained water and oil		1000	
fuel for 10,000 miles		1600	
total, including fuel			3400

CHAPTER 20
PUMPS

20.1 Fundamentals

In marine propulsion and auxiliary systems, it is necessary to move liquids from low-pressure areas to high-pressure areas and from low elevations to higher ones. In a steam plant feedwater must be moved from the deaerating feed tank, where the pressure may be about 50 psia, to the boiler, where the pressure may be 800 psig or more. Lubricating oil must be moved from the reduction gear sump to the gravity tank. Both locations are at atmospheric pressure, but the gravity tank may be 40 feet or more above the sump. A pump increases the pressure of the liquid so that it can flow to the desired destination. Fig. 20.1 shows a typical situation in which a pump moves a liquid from one tank to another.

Many different types of pumps have been developed. The best type for a given application depends on the required pressure increase, the flowrate, the nature of the liquid, and the vertical location of the pump with respect to the source and the destination of the fluid. Lubricating-oil pumps are very different from water pumps because oil is much more viscous than water. The main feed pump in a steam plant is very different from the pump that circulates seawater through the main condenser. This is because there is need for a large pressure rise in the feed pump, while the circulating pump must deliver a large mass flow of water but with only a small pressure rise.

In every pump, energy is transferred from the prime mover, which drives the pump, into the fluid. This prime mover is typically a steam turbine or an electric motor. The increase in fluid energy is displayed as an increase in pressure of the liquid.

Air and other gases must also be pumped in marine systems, but because the properties of a gas are so different from the properties of a liquid, the equipment used is different and is usually classed as a blower, fan, or compressor rather than a pump. **Blowers** and **fans** are used to pump gases where low pressures are involved. Combustion air is forced into a boiler furnace at a pressure just above atmospheric by a forced-draft fan. **Compressors** are used when higher pressures are needed. Compressors supply air to operate control systems, start diesel engines, and operate a variety of accessory equipment. Refrigeration compressors are used in refrigeration and air-conditioning systems.

20.2 Types of pumps

Two broad categories of pumps cover almost all types. **Positive-displacement pumps** function by alternately filling and emptying pumping chambers of fixed volume. The simplest such pump has a piston that moves up and down in a cylinder. Liquid flows into the cylinder as the piston moves away from the cylinder head, and out of the cylinder as the piston moves back toward the head. The pressure of the liquid is raised as the piston squeezes it against the head. **Kinetic pumps** increase the kinetic energy of the liquid first and then convert this kinetic energy into increased pressure. The most common types of pumps are discussed in detail in the following sections.

20.3 Centrifugal pumps

By far the most common type of kinetic pump is the **centrifugal pump**. The essential moving part of a centrifugal pump is called an **impeller**. Fig. 20.2 shows several impeller types. The impeller is mounted on a shaft, and the impeller and shaft are rotated by a prime mover such as an electric motor or a steam turbine. Liquid enters the impeller near its center through an open area called the **eye.** In Fig. 20.2 (a) the eye is the open area at the top. As this impeller rotates counter-clockwise, liquid is pushed outward by

centrifugal force. The impeller **vanes** force the liquid to rotate. Curved vanes can be clearly seen in Fig. 20.2 (b).

To understand how a centrifugal pump works, consider that the circumference of a circle increases as the radius increases. All points on the impeller must make one circular revolution in the same amount of time. Therefore, points at the outer rim must travel faster than points near the eye because they must travel farther in a given amount of time. The vanes force the liquid to rotate with the impeller. Thus, the liquid moves faster as it goes outward from eye to outer rim. This faster movement of the liquid increases its kinetic energy.

As the liquid leaves the impeller at the outer rim, it enters a carefully shaped portion of the pump casing called the **volute**. Fig. 20.3 shows a cross-section view of a pump, showing an impeller in a volute. As it flows through the volute the liquid slows down, which decreases its kinetic energy. The principle of energy conservation requires that some other component of its total energy must increase. Pressure divided by density is a form of energy called **head**, and this is the energy component that increases. Since the density of a liquid does not change, it is useful to think of pressure itself as a form of energy. The overall effect is that the liquid entering a pump gains kinetic energy in the impeller, and in the volute, this kinetic energy is converted to increased pressure.

Fig. 20.4 shows the construction of a typical small motor-driven centrifugal pump. An electric motor at the left turns the shaft (item 6). The impeller (item 2) is mounted on the right end of this shaft. Since there is a single shaft for motor and pump, this is called a **close-coupled** pump. Liquid enters the pump at the right through the threaded opening in the center of the suction cover (item 9). The discharge opening is not shown in this drawing.

Although the impeller does not actually touch the casing, the gap is small, and the casing can be eroded by the liquid. This erosion allows more of the liquid to recirculate through the small gap, reducing the performance of the pump. For this reason, many pumps have stationary **wearing rings**, items 25 and 27 in Fig. 20.4, recessed into the casing near the fast-spinning impeller. Because the wearing rings are removable, they can be replaced when worn to restore the performance of the pump.

From the motor, the shaft must pass through the pump casing to connect to the impeller. Where the shaft passes through the pump casing a seal prevents liquid from escaping. This seal must allow rapid rotation of the shaft but prevent liquid from escaping. These functions are accomplished with rings of **packing** (item 13). The exact nature of the packing depends on the temperature and pressure of the liquid being pumped. In simplest form, packing is a soft rope-like material that may be coated with Teflon or graphite to reduce friction as it rubs against the shaft. Lengths of packing are wrapped around the shaft and pushed into a chamber called a **stuffing box**. A metal **gland** (item 17) presses against the packing. As the packing is squeezed in the axial direction, it expands until it presses against the shaft to prevent leakage of liquid. Some of the liquid being pumped helps to lubricate the contact area between packing and shaft. A metal spacer called a **lantern ring** (item 29) is placed between rings of packing to allow this liquid to enter.

Eventually the pressure of the packing against the rotating shaft causes wear. A removable **sleeve** (item 14) is fitted around the shaft to protect it from damage. When the sleeve becomes worn it can be replaced.

Where large pressure increases are required pumps may have several impellers mounted on the same shaft in the same casing. The liquid flows from the discharge of one impeller, through a volute, to the eye of the next impeller. Thus each impeller contributes part of the necessary energy increase. Fig. 20.5 shows a boiler feed pump with four impellers. This pump would be driven by a steam turbine. It is not close-coupled, since the turbine shaft and the pump shaft are separate components connected by a

coupling.

Centrifugal pumps work best with liquids such as water that are not very viscous. Depending on size and configuration, they are suitable for a very wide range of flow rates and pressure increases. However, centrifugal pumps work best when the source of liquid is located above the pump. Typical applications of centrifugal pumps include condensate pumps, boiler feed pumps, fire pumps, and most other freshwater and seawater pumps.

20.4 Reciprocating pumps

One of the oldest types of pumps is the steam-powered **reciprocating pump**. A typical example of this type is shown in Fig. 20.6. The upper part of this pump is the **steam end**. Steam can be admitted above or below the **steam piston** to move it down or up. To move the piston down, steam is admitted above the steam piston while steam exhausts from below this piston. As the piston reaches the bottom of its stroke, valves open and close to admit steam below the piston and to exhaust the steam from above it. Thus, the piston is made to move alternately up and down. The steam end is described as **double acting**, because steam can be admitted alternately above and below the piston.

The steam piston is attached to the **steam piston rod** (item 3) and this is joined to the **liquid piston rod** (item 7). Attached to the end of the liquid piston rod is the **liquid piston** (item 6). Because the two rods are joined, the liquid piston moves up and down with the steam piston. The liquid end of the pump is also double acting. As the liquid piston moves down, liquid is discharged from beneath the piston and drawn in above it. The reverse happens as the piston moves up.

Valves control the flow of steam and water. Water is drawn in at low pressure and discharged at high pressure. This type of pump is capable of high discharge pressures, but it is limited to low flow rates. A reciprocating pump is a good choice when the source of liquid is below the pump. These pumps are often used to pump water out of the bilge or to **strip** the last of the cargo oil from the bottoms of cargo tanks on tankers.

20.5 Rotary pumps

Another common type of positive displacement pump is the rotary pump. **Gear pumps** and **screw pumps** are included in this category. Fig. 20.7 shows an external gear pump. The shaft carrying one of the gears is connected to an electric motor or other prime mover. As this gear rotates, it drives the other gear through direct contact. In the figure, the upper gear rotates clockwise, turning the lower gear counterclockwise. Liquid enters through the suction opening at the left side. Pockets of liquid become trapped against the pump casing. As the gears rotate, the fluid is carried around a half circle by rotation. Because the two gears mesh at the center of the pump, the liquid cannot continue around the full circle. Instead, it is forced out through the discharge opening. The liquid being pumped also lubricates the point of contact between the two gears. Gear pumps are commonly used to pump viscous liquids such as fuel and lubricating oil, but only when low flow rates are required. These pumps can lift liquids from sources below the pump, and they can produce high pressures at the discharge.

Where higher flow rates of viscous liquids are needed, screw pumps are often used. Fig. 20.8 shows a typical twin-screw pump. Liquid enters this pump from below, and it flows to the ends of the rotating screws. The screws are similar to common machine screws, but the "threads" are cut much deeper into the shaft. Liquid becomes trapped in the space between the threads, enclosed by the casing. As the screws rotate, the liquid in these spaces moves toward the middle of the pump. Upon reaching the center, the liquid has nowhere to go but out through the discharge opening at the top. Note that liquid is pulled toward the middle from the left and right ends of the pump. The upper screw shaft extends through the casing where it is connected to a prime mover such as an electric motor. The upper screw is called the **power screw**, and the lower is called the **idler screw**. Just inside the casing on the right side, there is a gear on each shaft. These gears allow the upper shaft to drive the lower one. These gears are called **timing gears**. The pumping screws are separated by very small clearances, but they do not actually come into contact.

Not all screw pumps have timing gears. Those that do are called **timed pumps**, and those that do not are called **untimed pumps**. In untimed pumps the idler screw is driven through direct contact with the power screw, with lubrication provided by the liquid being pumped. Untimed pumps are therefore suitable only for clean liquids such as lubricating oil, while timed pumps are necessary for aggressive liquids like heavy fuel oil.

20.6 Jet pumps

Jet pumps can be used to move liquids and gases.

Jet pumps are used as **air ejectors**, to remove air from the vacuum condensers of steam plants. Ejectors may also be used to remove air from the shells of distilling plants. A typical ejector is shown in Fig. 20.9. Steam at about 150 psig is admitted to a nozzle, where it accelerates to very high speed. The high-speed steam jet draws the air from the suction chamber into the diffuser tube and propels it toward the discharge. This creates a partial vacuum in the suction inlet area, and more air is drawn from the source. The diffuser acts like the volute of a centrifugal pump to convert the high kinetic energy of the steam-air mixture into increased pressure.

Many condensers, including main condensers on steam ships, operate at a very high vacuum (that is, at very low pressure), and it is necessary to use two air ejectors in series to raise the pressure of the air up to one atmosphere. These **two-stage** air ejectors usually have a small condenser after each stage to condense the steam that was used in that stage. The condenser after the first stage is called the inter-condenser, and the condenser after the second stage is called the aftercondenser. Drains from the inter-condenser flow through a **loop seal** to the main condenser, while drains from the aftercondenser flow to the atmospheric drain tank.

Jet pumps are also used as **eductors**, where pressurized liquid, instead of steam, is used as the propelling fluid. Because eductors can lift liquid from a source below the pump, they are often used with seawater as the propelling fluid to pump water from bilges or to pump brine overboard from a distilling plant. On some tankers, eductors using cargo oil from the discharge of the cargo pumps are used to **strip** the last of the cargo oil from the bottoms of cargo tanks.

20.7 Liquid-ring vacuum pumps

Fig. 20.10 is a schematic drawing of a **liquid-ring vacuum pump**. The circular rotor, item 3, rotates inside the elliptical casing, item 5. The casing is partly filled with water, which remains in the pump, but which is spun by the rotor, forming a "liquid ring" that follows the elliptical shape of the casing. Because the rotor is circular, but the casing is elliptical, the water moves in and out along the rotor vanes as the rotor turns. In this way, the water between any two vanes acts like a piston. In Fig. 20.10, following the rotor clockwise from point 3, where the water fills the vane space, as the water follows the shape of the casing air is drawn in through the inlet port. From point 4 to point 5 the air is compressed, and it is then forced out through the discharge port. By point 6 the water again fills the vane space, and the process repeats through point 7 back to point 1. Thus there are two inlet-discharge cycles per revolution.

Liquid-ring vacuum pumps are often used instead of air ejectors to remove air from condensers.

20.8 Blowers and fans

Blowers and **fans** are used to supply large amounts of air at relatively low pressure. Although the distinction is somewhat arbitrary, the term fan is usually used when the pressure rise is 1 psi or less. The term blower is used where the discharge pressure is up to 40 psig. Machines that produce higher pressures are usually called **compressors**.

Fig. 20.11 shows a typical axial-flow blower. Air enters from the atmosphere at the left. It passes through two sets of rotating blades (item 2) that are similar to turbine blades, and then flows through the **diffuser** into a large discharge chamber. The rotating blades are mounted on a shaft (item 8) that is driven by a prime mover. Following each set of **rotating blades** is a set of **guide vanes** (item 3) that remove the swirling motion from the flow. In this type of blower, the kinetic energy of the air increases in the rotating blades, and is converted to pressure in the guide vanes and in the diffuser.

20.9 Compressors

When gas pressures above 40 psig are required, a compressor is used. Many compressors are of the reciprocating-piston type like the one shown in Fig. 20.12. An electric motor drives the **flywheel** at the left through a group of belts, and the flywheel turns the **crankshaft** (item 4). The cranks that give the crankshaft its characteristic shape cause the pistons (item 3) to move up and down as the crankshaft rotates. As the pistons move up and down, air is drawn in and forced out like the liquid in a reciprocating pump. The two pistons operate in series: the air is partly compressed in the first cylinder and then compressed further in the second cylinder. Compressing air causes its temperature to rise, so the air passes through an **intercooler** between the first cylinder and the second. A fan blows air over the intercooler tubes, and heat is transferred from the partially compressed air inside the tubes to atmospheric air flowing outside those tubes. Intercooler tubes usually have fins on them to increase the heat transfer surface. In some compressors the intercooler uses seawater rather than atmospheric air to absorb heat from the compressed air.

Rotary compressors are also used. Some rotary compressors resemble screw-type pumps. Fig. 20.13 shows one of these. Often oil or water is sprayed into these compressors to provide lubrication where the screws touch and to help seal the small clearances between the rotors. Timing gears are sometimes used to prevent contact between the rotating screws. Where timing gears are used, it may be unnecessary to spray oil or water into the compressor. This is desirable in applications where the presence of oil or water droplets in the compressed air would cause problems.

Compressors are also used in refrigeration and air conditioning systems. In these cases a refrigerant vapor flows through the machine instead of air. Refrigeration compressors differ from air compressors in the details, but the basic concepts and configuration are the same.

20.10 Pump operation

The piston or rotor of a positive-displacement pump cannot move unless the suction and discharge valves for the liquid are open. These valves <u>must</u> be opened before the prime mover is started. In some cases serious damage can result if this procedure is not followed.

Conversely, liquid cannot flow through a positive displacement pump unless the piston or rotor is moving. Often the valves are left open even when the pump is not in use. If the suction and discharge valves are always open, the pump can be placed in service immediately just by starting the prime mover. Motor-driven positive-displacement pumps may be fitted in parallel to provide lubricating oil to a steam turbine and reduction gear, or to a diesel engine. One of these pumps is always running, and the other is on standby. Should the pump in service fail, an automatic device starts the motor of the standby pump. In this way there should be no interruption in the flow of oil.

In contrast to positive-displacement pumps, centrifugal-pump impellers can rotate whether or not the suction and discharge valves are open. Centrifugal pumps are sometimes started with the suction valve open and the discharge valve closed, but they must not be operated for extended periods in this condition. Friction against the rotating impeller tends to heat up the liquid. Unless the heated liquid flows out of the pump, it can reach its boiling point. If the liquid boils, serious damage can occur.

Liquid can flow through a centrifugal pump when the impeller is not rotating. This means that where two centrifugal pumps are installed in series, liquid will flow even if one of them is not running. It also means that where pumps are installed in parallel, liquid can flow backward through the idle pump. This reverse flow can be prevented by installing a non-return valve at the discharge of each pump.

Sometimes centrifugal pumps are located above the source of liquid, but additional equipment is needed in this case. The bottom end of the **suction (inlet) pipe** must be fitted with a **foot valve**. This valve is a non-return valve that allows liquid to flow toward the pump but not back to the source. In some cases, the pump casing is fitted with a **vent valve** to let air out, and a **priming connection** to let liquid in from a source higher than the pump. Before the pump can be started, liquid must be admitted through the priming connection to fill the suction line and the pump casing. The displaced air escapes through the vent, and the foot valve prevents the liquid from draining into the source tank. Once the suction piping and pump casing are full of liquid, the pump can be successfully started. Alternatively, a vacuum priming device may be used to prime the pump.

20.11 Starting procedure for a centrifugal pump

1. Make sure that the pump is of the centrifugal type.
2. Open the suction valve, and close the discharge valve.
3. If the pump is fitted with a priming connection, open the vent valve on the pump casing and open the inlet valve on the priming line. Allow priming liquid to flow into the pump until a solid stream exits from the vent.
4. Close the vent valve first, then the inlet valve on the priming line.
5. Start the motor or turbine that drives the pump.
6. Gradually open the pump discharge valve. Be sure to open it all the way.
7. Verify that liquid is being discharged from the pump.

20.12 Starting procedure for a positive-displacement pump

1. Make sure that the pump is of the positive-displacement type.
2. Open the suction and discharge valves. Be aware that these valves may already be open.
3. Start the prime mover of the pump. If the pump is driven by an electric motor, close the breaker for the motor. If the pump is driven by a reciprocating steam engine, open the inlet and exhaust valves on the steam end. Be aware that the exhaust valve may already be open.
4. Verify that the pump is operating smoothly.

CHAPTER 21
PIPING COMPONENTS

21.1 Introduction

Piping systems are used to convey fluids. The term "piping" generally refers to the pipe, valves, fittings, flanges, and other components that form the system.

21.2 Pipe and tubing

There is no definite rule for distinguishing between **pipe** and **tubing**, and accepted terminology is established by usage. One difference is the wall thickness: typically, pipe has heavier walls. Pipe generally conveys fluid flow from one location to another, whereas tubing may direct static pressures for control or measurement. The term tubing also applies to the cylindrical elements that separate fluids in boilers and other heat exchangers.

Pipe and tubing are also distinguished by the way in which segments of a system are formed and joined. Typically, tubing is bent using hand tools while piping is bent using pipe-bending machines. Tubing is joined by **soldering**, **flaring**, **swaging**, or by **compression fittings**. Piping is joined by **threading**, by **welding**, or by **bolted flanges**.

21.2.1 Materials and class

The materials used in piping systems and valves must be compatible with the fluid conveyed and with each other. Common materials are carbon steel, stainless steel, chrome-molybdenum steel, cast iron, copper, brass, bronze, copper nickel, and plastic. Selection of the material is based on the fluid type and service conditions. ASTM A-53 and ASTM A106, Grade B carbon steel pipes are commonly used on board ship for steam, condensate, oil, and air systems. They may be used for seawater systems, but where corrosion is a concern, copper-nickel or other brass or bronze alloys are preferred. Copper tubing may be used on low-pressure, low-temperature lines, such as gage sensing lines.

Materials joined in a piping system must be similar, to avoid **galvanic action** (see Table 21.1). For example, mixing steel and brass fittings in a seawater line would result in accelerated galvanic corrosion of the steel, as it acts like a **sacrificial anode**. The less active that the material is on the galvanic-series table (or more noble the material), the more resistant it is to galvanic attack. This table also indicates why zinc is commonly used as a sacrificial anode for the protection of metals exposed to seawater. For a summary of typical piping materials used for various shipboard systems refer to Table 21.2.

Shipboard piping is divided into two categories, **Class I** and **Class II** (or Group I and II), ranking the piping with regard to the fluid energy and hazard potential. Typical limits for working pressure and temperature are shown in Table 21.3: for service at or above the pressure or temperature cited, the piping is designated as Class I (or Group I) and must meet higher standards for strength and safety. Class II includes all other piping including exhaust-gas piping. The values given in the table are typical, but will differ from one regulatory agency to another.

As indicated by the pipe class, pipe and fitting strength is affected by temperature as well as pressure, and both must be considered in the selection of materials and wall thickness. As the temperature increases material strength decreases. Higher pressure results in higher stresses on components. Heavier pipe and fittings are required to meet strength requirements as system pressures and/or temperatures increase. At low temperatures, some materials become brittle and may be susceptible to sudden and catastrophic failure.

21.2.2 Sizing

Pipe is sized by a **nominal pipe size (NPS)**, which is in general use worldwide. Standard pipe sizes range from 1/8 inch to 36 inches. The nominal pipe size is based on a *nominal* inside diameter for sizes up to 12 inches diameter. Since the pipe is joined using standard-size fittings, the outside diameter must be the same regardless of wall thickness. Consequently, pipe-wall thickness is achieved by adjusting the inside diameter. Standard **schedules** are used to define wall thickness, and the relationship of schedule number to wall thickness varies according to pipe size. Tables must be consulted in selecting pipe. Table 21.4 shows some sample dimensions. The larger the schedule number, the thicker the pipe wall. For example, a 1-inch NPS pipe has an inside diameter of 1 inch *in name only*: for 1-inch Schedule 40 pipe the wall is 0.133 inches thick and the ID is 1.049 inches, but for 1-inch Schedule 80 pipe the wall is 0.179 inches and the ID is 0.957 inches. But in all cases of 1-inch NPS, the OD is 1.375 inches. In smaller sizes the nominal pipe size (nominal ID) is not very close to the actual ID. As an example, 1/8 inch NPS, Schedule 40 pipe has an ID of 0.269, more than twice the nominal value. For pipe sizes of 14 inches and larger, the nominal pipe size is based on the outside diameter. In international practice the same dimensions are used, but the nominal pipe size is designated in rounded millimeters, calculated at 25 mm to the inch. For example a 150-mm pipe is 6-inch NPS.

A pipe-wall thickness convention sometimes used is the **Iron Pipe Size (IPS)** system. The IPS system uses the terminology Standard (Std), Extra Strong (XS), and Double Extra Strong (XXS). It is evident in Table 21.4 that Std corresponds to Schedule 40 and XS to Schedule 80 but only for pipe sizes through 8-inch NPS.

Pipe can either be **seamed** or **seamless**. Seamed pipe is formed by rolling a flat plate into a cylindrical shape and then welding along its longitudinal seam. The outside weld bead is subsequently ground down flush with the outside diameter of the pipe so that it can be joined using standard threaded or slip-on fittings. In seamed pipe the inside weld bead can be observed as a raised surface when looking into the pipe. Seamless pipe is formed by a drawing process where red-hot metal is drawn over a piercing mandrel, forming the tubular shape directly. Seamless pipe is considered to be superior to seamed pipe and is used as a general-purpose pipe. Seamless pipe is required for all fuel-oil service piping, for steam lines above 450°F, and all service pressures above 350 psig.

Tubing is specified by its outside diameter and decimal wall thickness, where the wall thickness is usually specified by gage sizes. A common tubing gage is the **Birmingham Wire Gage** or **BWG**. Some examples of BWG sizes are shown in Table 21.5. Note that as the BWG gage number *increases*, the wall thickness *decreases*. Using the above information, if a boiler has 2"-#8 BWG screen tubes, the outside diameter of each tube is 2 inches and the wall thickness is 0.165 inches. Consequently, the ID of the tube is: 2" - (2 x 0.165") = 1.770 inches. In contrast to pipe schedule designations, the BWG value does not vary with tubing size.

21.2.3 Joining methods

Pipe is joined together and to equipment using fitting connections that are threaded, flanged, welded, brazed or soldered, or with compression fittings. See Fig. 21.1 for illustrations of common pipe fittings.

The most common method for joining pipe is the threaded (or screwed) connection. Threaded joints are used for **small-bore piping**, even at high pressures. To maintain pipe strength when using threaded joints, it is necessary that sufficient pipe wall thickness is available to compensate for the loss of outside diameter from cutting the threads into the pipe wall. Common practice is to specify Schedule 80 pipe for small-bore, high-pressure pipe when the joining method is by threading.

Pipe threads differ from machine threads in that they are cut on a taper, i.e., the thread diameter is smaller at the end of the pipe and is progressively larger along the pipe length. As a result of the taper, when the pipe is screwed into a fitting, the threads force an increasing interference that assists in sealing against leakage. Compounds are normally applied to the external threads to fill any imperfections in the interstices between the pipe and fitting threads. Teflon tape or Teflon-bearing pipe dope is often used on lower-temperature connections, and lubricants bearing copper or silver metal flakes are used in higher-pressure, higher-temperature joints. In addition to helping seal the threads against leakage, these compounds prevent **galling** of the threads and seizing, thereby facilitating future disassembly for maintenance or repair.

Union or flange connections are used at desirable break points for disconnecting piping from equipment during repairs. Flanges consist of circular flat-faced disks that are bolted together, with a compliant gasket material installed between them. This type of connection provides a means for fabricating pipe runs in manageable sections for field assembly without welding, and provides the necessary break points for maintenance and repair. For high-pressure piping, it is standard practice to use **raised-face flanges**, where the flange face in the area within the bolt holes is raised to increase compression of the gasket in the sealing area. It should be noted that in any installation using cast-iron valves, raised faces should not be used on the flanges, and if present, they should be machined off. Using flat faces with cast iron avoids high bending stresses on the cast-iron flange.

Flanges can be attached to pipe by threading or, more commonly, by welding. Welded flanges may be of the **slip-on** type, where the flange disk is slipped over the pipe, with the flange face protruding slightly past the pipe end. The flange is then fillet welded inside and out. **Weld-neck flanges** are forged with a beveled extension or neck attached to the flange for butt-welding to pipe, and are used for high pressure systems.

Welded joints are used for general-purpose piping and are required for high-pressure, high-temperature piping systems. As with welded flanges, two types of welded fittings are used, **socket-welded fittings** and **butt-welded fittings**. The socket-welded fittings are used for small-bore piping. The fitting is manufactured so that the pipe is inserted into the fitting, after which a fillet weld is made between the outside edge of the fitting and the pipe junction. Socket-welded fittings are generally acceptable up to 2-inches NPS.

Larger-bore pipe systems are made-up using butt-welded fittings and joints. The welded joint is prepared so that the fitting and pipe are both beveled and then joined with a **full-penetration weld** which develops a continuous weld bead on the inside diameter of the pipe as well as the outside. It may be necessary to machine the inside diameter of fittings so that pipe-wall-to-fitting thicknesses match before welding, to reduce stress concentrations. Sometimes **backing rings** are used. Backing rings are thin metal rings that slip into the pipe to keep the fitting-to-pipe alignment during welding. The rings are fused to the weld during the joining process, but they may provide a location for crevice corrosion. Butt-welded joints are conducive to **radiographic** or **x-ray testing** for high-reliability applications.

21.2.4 Gaskets

Common **gasket** materials include rubber, paper, cork, and metal. Rubber comes in various hardnesses and is often reinforced with wire or cloth to prevent extrusion from the flange during bolting. Some gasketed joints use rubber O-rings to obtain the seal. The gasket material must be selected to withstand the system temperatures and pressures, and must chemically compatible with the contained fluid. Paper and cork gaskets are commonly found on oil systems. It is good practice to coat paper gaskets with a thin layer of the oil before assembly. Other services may use gasket-sealing compounds.

Metal gaskets are used in high-pressure, high-temperature applications. Copper-ring gaskets are often found within valve bonnets of high-pressure steam systems, and should be annealed when reused after valve disassembly. The spiral-wound metal gasket illustrated in Fig. 21.2 is another commonly used type. Pressure acts against the open end of the gasket's V shape, tending to spread the V shape against the raised-face flanges, maintaining a tight seal. A reinforcing ring prevents gasket blowout and also locates the gasket concentrically within the flange bolt pattern. When spiral-wound gaskets are used within valve bonnets, the valve body has a machined recess to restrain the gasket, and the metal retaining ring is not needed. Spiral-wound gaskets are always used with raised-face sealing surfaces.

21.2.5 Piping support

In piping installation, a horizontal pipe is referred to as a **run** and a vertical pipe as a **riser** (see Fig. 21.3). In routing pipe, it is important to plan the installation to avoid over-stressing either the piping system or the attached equipment. To reduce stresses induced by **thermal expansion**, bends and loops are used as in Fig. 21.4. As the pipe expands the system flexes to absorb the movement. Anchor points are provided by connections to equipment and to rigid supports, and are used to control thermal expansion in desired directions, and to prevent **water-hammer** damage to the piping system. Water hammer is a steam-pipe phenomenon where condensate is accelerated by high-velocity steam. When the condensate strikes a fitting or valve the sudden change in direction produces an effect similar to the blow of a hammer.

Piping supports or **hangers** are shown in Fig. 21.5. For large diameter, high-temperature systems, it is common to use **roller supports** with insulation shields. The roller allows free axial-expansion movement to minimize thermal stresses. **Spring supports** are fitted near risers so that upper and lower supports equally share the pipe weights over a wide range of operating temperatures. For horizontal runs, simple clevis-type supports are often adequate. Long clevis rods allow thermal expansion and contraction by moving from the vertical.

21.3 Valves

21.3.1 Classification

Valves are the devices used to regulate or stop fluid flow. The major types are **globe**, **gate**, **butterfly**, **ball**, and **plug valves,** illustrated in Figs. 21.6 and 21.7. Other types are **check valves** that permit flow in one direction only, and **relief** and **safety valves**, which protect systems from over-pressurization.

Valves may be classified by their function or service as **stop**, **check**, **stop-check**, **relief**, **safety**, **regulating**, **pressure-reducing**, and **back-pressure regulating**.

Stop valves are used to isolate equipment and may be of the globe, gate, butterfly, ball, or plug type. Although globe and butterfly valves may be throttled to regulate flow, when used as stop valves they are either fully open or fully closed.

Check valves are of the swing-check (clapper valves) or of the lift-check type. In the swing-check valve, a disk pivots on a hinged joint located above the disk's center of gravity. Some larger swing-check valves, such as scoop-injection valves, may have a weighted lever extending through the valve body to balance the valve-disk weight and reduce the actuation pressure requirements. The lever can also be used to observe free operation. Check valves are mounted horizontally with the lifting disk closed by gravity and by reverse flow. A variation of the check-valve construction often used in hydraulic systems uses a ball-shaped disk held against a seat by a spring. Flow occurs when pressure acting underneath the ball lifts the ball from the seat against the spring. This type of check valve can be mounted horizontally or vertically.

Stop-check valves are stop valves where the disk is not mechanically linked to the stem, but is free to slide along the stem. In the closed position, the stem holds the disk against the seat. In the open position, the disk behaves like a lift-check valve with the tip of the valve stem acting as a guide to maintain disk alignment to the seat. Like the stop valve, the stop-check can be used to throttle flow. Stop-check valves are common for boiler feedwater and steam stop valves, bilge and ballast suction manifolds, pump discharge lines, and other applications where reverse flow must be avoided.

Relief and safety valves are used to prevent over-pressurization of piping and/or equipment (see Fig. 21.7). The valves are similar to angle globe valves in that a disk seals against a seat, but with the disk held in place by a spring. Pressure acts on the underside of the disk producing a lifting force. When the lifting force exceeds the spring's compressive force, the valve lifts and relieves excess pressure. The valves are fitted with provision for adjusting the spring force and consequently the relief setting.

Relief valves differ from safety valves in that the higher the pressure, the more the disk lifts off its seat. Therefore relieving capacity increases as over-pressurization increases. On the other hand, the safety valve is designed to stay completely closed until the popping pressure is reached. At that point, the valve is designed to lift sharply and stay wide open until the required **blowdown,** or reduction in pressure (typically 10%), is reached. Consequently, the safety valve has sudden and positive lifting and reseating pressures. Safety valves are either of the **nozzle-reaction** or **huddling-chamber** type, the names describing the techniques employed to obtain the sudden popping and blowdown. Safety valves are used on boiler steam drums and superheater outlets.

21.3.2 Valve selection

Considerations in selecting valves include the fluid conveyed, material compatibility, the function of the valve, the desired reliability, weight and size restrictions, and cost. The valve must be matched to the pressure and temperature of the fluid. As with pipe, as the temperature of the fluid rises, the strength of the valve material goes down. Consequently a pressure-temperature relationship exists in selecting valves. In the U.S., valves are classified by a **pound (#) Class** which represents more than just pressure, but accounts for temperature as well. Standard valve classes include:

Cast Iron: 125#, 250# Classes,
Forged Steel: 150# , 300#, 400# , 600# , 1200#, and 1500# Classes.

Table 21.6 shows the pressure-temperature relationship for a valve constructed of ASTM A217 Grade WC9:3 (forged steel).

It should be noted that when flanged cast iron valves are selected, the mating flanges must not have raised faces: mating faces must be flush to avoid unacceptable tensile stresses caused by bending of the cast iron flange as the valve is bolted in place.

It is common to see the valve size, the valve class, a flow direction arrow, and the letters **WOG** cast into the body of a valve. WOG means that the valve is adequate for water, oil, or gas service.

21.3.3 Valve construction

The major components of a valve are the disk and seat, the body and bonnet, stem, packing gland, packing, and handwheel. See Figs. 21.6 and 21.7.

The globe valve uses a flat or tapered disk that is raised and lowered by a valve stem. In the closed position, the disk is screwed down tightly against the valve seat thereby stopping flow. Globe valves can have a straight-through or angle configuration. In the straight globe valve, the inlet and outlet are located in line with each other. The angle valve (Fig. 21.7b) diverts flow through 90°, functionally combining the globe valve with an elbow. Straight valves can have a Y-configuration where the stem, disk, and seat are oriented at 45° within the body, reducing pressure losses. Globe valves are usually installed with upstream pressure acting under the disk. Globe valves are excellent stop valves and can be conveniently refurbished by lapping the disk and seat with grinding compounds. Globe valves are also good for throttling in the nearly-closed position, especially when mostly closed, but they tend reach full flow rather quickly. A variation of the globe valve is the needle valve, which uses a sharply tapered plug in lieu of a disk, and is designed to provide very fine flow control, especially at low flow rates.

The gate valve (Figs. 21.6a, b, and c) consists of a gate-like disk, perpendicular to the direction of flow, which slides in guides in the valve body. The gate is wedge-shaped and is forced down into a matching tapered seat. When the gate is lifted flow is straight through the valve body and flow resistance is low. Gate valves can be of the rising-stem type where the stem is linked into the disk and it pulls the gate open as it rises; or gate valves can be of the non-rising stem type where the stem is threaded into the disk and the disk travels up the thread as the stem is rotated. Rising-stem valves inherently indicate the valve position by the height of the stem, while non-rising stem valves are non-indicating unless fitted with a threaded position indicator. The non-rising stem valve takes less space.

The ball valve (Fig. 21.7c) consists of a polished ball with a hole that aligns with the flow when the valve is open. Actuation is by a lever that rotates the ball through 90°. When the handle is parallel to the pipe the valve is fully open, and when the handle is at 90° the valve is fully closed. The ball is sealed in the body by rubber or plastic rings. Ball valves are generally designed for open/closed operation, although they throttle well. Advantages of ball valves are quick closing by quarter-turn actuation, highly reliable, and cause little pressure loss. Ball valves are manufactured with “reduced ports” or “full ports” referring to the ball-hole diameter relative to the pipe inside diameter. The valves are assembled as either top entry or side entry. Ball valves tend to be used for smaller sizes as large sizes are quite heavy and require geared actuators to operate.

The plug valve is similar to the ball valve, except that it uses a tapered plug with a slotted hole instead of a ball. The plug valve uses substantial downward pressure of the tapered plug into to body to obtain its seal. For this reason, plug valves require very high actuation forces to operate. It is not common to find plug valves on ships.

The butterfly valve (Fig. 21.7d) has a flat disk with a lever handle attached to a rotating stem. Like the ball valve, actuation is by 90° rotation of the stem handle. When the handle is perpendicular to the piping the disk blocks the flow, sealing against a seat that is usually coated with an elastomeric material. When the handle is in-line with the pipe the valve is open. Since the disk is always in the flow, there is an associated pressure drop, but because of the straight-through flow it is usually moderate. Advantages of butterfly valves are quick quarter-turn actuation, relatively low cost, light weight, compact dimensions, and the ability to be throttled. Because of sealing problems when closed, butterfly valves are not usually found in high-pressure applications. Butterfly valves are common in seawater systems and in tanker cargo-oil systems. High-performance butterfly valves are a refinement that use disks mounted eccentrically on the stem and cam-like motion to seal very tightly against its seat. These valves are more reliable, seal tighter, and can be used in higher-pressure applications than the conventional butterfly valve.

21.3.4 Valve actuators

The simplest actuators are handwheels for gate or globe valves and lever handles for butterfly, ball, or plug valves. Larger valves require more force to operate. On gate and globe valves, larger handwheels are used and sometimes **slammer** devices are built into the handwheels to provide high impact forces. For horizontally mounted overhead valves, **chain and sprocket actuators** may be used, and for other difficult-to-reach valves, flexible shafts or **reach rods** using universal joints may be installed. If the force to operate becomes very high, it is common to see **geared actuators** to provide mechanical advantage. Some valves are fitted with electric, pneumatic, or hydraulic motors. These actuators provide the required operating forces and also permit remote or automatic actuation.

21.3.5 Regulating valves

Regulating valves, also called control valves, are used to automatically control the flow of fluid in response to level, pressure, or temperature. An example is the self-contained and self-actuated pressure-reducing valve of Fig. 21.8. In this self-contained valve, internal passages apply system pressure to the controlling **diaphragm** to create the actuation force. The resulting diaphragm force is opposed by the adjusting spring force. The spring force balances the diaphragm force when the regulator has reached its set position. Adjustments to the spring compression change the regulator setting. Similar valves are used as self-actuated temperature regulators, with an external bulb containing a volatile fluid is used to sense temperature and transmit a corresponding pressure signal through a capillary tube to the valve's diaphragm. In some applications **bellows** are used instead of diaphragms.

Other regulating valves are pneumatically actuated using air signals to control the regulator position. An example is steam-pressure reducing valve of Fig. 21.9. The pneumatic **pilot controller** on the left, which is supplied with control air at constant pressure, senses the controlled steam pressure downstream of the regulating valve, and generates a signal air pressure that is proportional to the steam pressure. This signal pressure, typically 3 to 15 psig (0.2 to 1 bar), is directed to the diaphragm in the **actuator** section of the regulating valve. The resulting diaphragm force is opposed by the actuator spring. The regulator setting is adjusted at the pilot controller.

A relief valve is always installed on the low-pressure side of a pressure-reducing valve to avoid damage in case the reducing valve malfunctions. Common practice is to install stop valves on the inlet and outlet of the regulating valve, with a manually operated valve in a bypass to enable the regulator to be repaired while the system is kept in service. Bypass valves are usually of the globe type so that they may be throttled for good control.

In addition to pressure-reducing valves, there are back-pressure regulating valves. A pressure-reducing valve receives high-pressure fluid from a source and reduces it to a lower pressure for use elsewhere. A pressure-reducing valve therefore regulates *downstream* pressure. On the other hand, a back-pressure regulating valve regulates *upstream* pressure by dumping fluid in a controlled manner to another part of the system.

A common use of pressure regulators and back-pressure regulators is in the auxiliary exhaust line (or maintained exhaust line) of a steam plant, which requires a steady pressure for reliable operation of the main feed pump. If the pressure in the line is too low a pressure-regulating valve supplies steam to maintain the pressure. If the pressure is too high a dump valve (a back-pressure regulating valve) will automatically open to dump steam to a condenser.

21.3.6 Steam traps

Steam traps (see Fig. 21.10) are devices used to automatically drain condensate from steam lines or heat exchangers. In steam lines, they are installed to avoid **water hammer** and corrosion. In a steam heaters or a heating-coil return line, a trap forces the steam to stay in the heater or coil until it gives up all of its latent heat and condenses. A malfunctioning trap reduces heat exchanger and heating coil performance and may also result in lost thermal energy.

Steam traps are of three basic types: **mechanical**, **thermostatic**, and **thermodynamic**. Examples of mechanical traps are the float type, the bucket type, and the inverted-bucket type. **Float traps** (Fig. 21.10f) use the greater density of condensate to create a buoyant force to lift a float and open a valve. The **bucket trap** (Fig. 21.10e) uses a bucket that is initially lifted by condensate, closing the valve in the trap. When enough condensate enters the trap, the bucket floods and sinks, causing the valve to open. Steam pressure then forces the condensate out of the bucket until the bucket becomes buoyant again, rising and closing the valve. Because bucket traps can trap air they are rarely used aboard ship. In **inverted-bucket traps** (Fig. 21.10a and b), as steam enters a flooded trap, it fills the bucket creating a buoyant force that lifts the bucket and closes the valve. To keep the inverted bucket from trapping air a vent is provided in the top of the bucket.

Thermostatic traps use a fluid-filled bellows (Fig. 21.10g) or a bi-metallic element (Fig. 21.10h) to actuate the trap. The fluid or metal expands as it absorbs heat when steam is present, thus closing the valve. When the steam in the trap condenses, the trap cools, and the valve opens. Some traps combine float trap features with those of thermostatic traps to allow high-volume flow with automatic venting.

Thermodynamic traps of the **thermodynamic disk** and **impulse** types (Fig. 21.10c and d) use the principle that as hot condensate passes through an orifice, it will drop in pressure and flash into steam. Impulse-pressure from the flashed steam acts on a larger surface area on the top of a disk, momentarily closing the valve. In normal operation, these traps continually open and close as hot condensate passes, and proper operation can be verified by a chattering noise using a stethoscope. A large volume of condensate approaching the trap will act on the underside of the disk, holding it open.

21.4 Basic valve maintenance

Problems with valves include external leakage, inability to close tightly, and binding of moving parts.

The most common problem experienced with valves is external leakage at the valve-stem **packing**. Unchecked leakage will damage the packing and maybe the valve stem. Frequent operation of the valve can cause packing to weep. Steam-valve packing tends to dry out over time and begin to leak. Gland leakage can be corrected by tightening the packing gland until the leakage stops. If the gland is taken-up all the way, another ring of packing can be added or the valve can be repacked.

Some valves are designed to be repacked under pressure, but most valves must be completely isolated from internal pressure before repacking. To repack a valve, the packing gland is backed all the way out and the packing completely removed using a packing puller and picks. All remnants of packing are removed so that new rings will seat fully and evenly down to the bottom of the gland. New packing is cut with a skive joint (where the packing ends are angled at 45° and placed into the gland so that the *short dimension* of the cut is seen from the top of the gland. Rings are inserted with a progressive 120° annular stagger so that fluid cannot leak from lined-up seams. The gland is tightened down sufficiently to prevent leakage from the gland, while still allowing the stem to be rotated with reasonable effort. If a valve gland leaks after it is placed in service, it is tightened down until the leakage stops. Note that this practice is contrary to pump-packing gland practice, where limited leakage is generally necessary for continuous cooling and lubrication.

Failure of a valve to close tightly may be caused by damage or by foreign matter trapped between the disk and seat. Before replacing or refurbishing, it is good practice to try unseating and reseating the valve to clear the seat.

Valves that are used for throttling are subject to **wire drawing** or cutting of the valve disk or seat because of erosion, allowing leakage through the valve when closed. Gate-valve disks and seats are difficult to repair and gate valves are therefore not normally used for throttling. Globe-valve disks and seats can be repaired if minor cutting or pitting of the disk or seat occurs by **lapping**. **Lapping compounds** comprise abrasive grit held in suspension by heavy cutting oil. With the valve disassembled, lapping compound is applied between the sealing surfaces of the disk and seat, and the disk is rotated back and forth under applied pressure, thus grinding the disk in place to match the seat and removing any blemishes. The valve disk is often mounted to a makeshift stem possibly attached to a speed wrench. As the disk is rotated back and forth, the angle of lapping rotation is periodically changed so that the valve is ground evenly around. As the disk and seat become smoother, the coarser lapping compound is wiped away and replaced with a finer grit. If care is taken to ensure that the disk is maintained straight against the seat, the valve can be refurbished to a like-new condition.

Sticking of valves can occur if they are jammed open or closed, especially when hot. To prevent jamming, valves should be closed only tightly enough to stop flow, and when opening a valve, the handwheel should always be closed a quarter of a turn. Sticking can also be caused by a packing gland that is taken-up too tightly, by a gland tightened unevenly, by a bent valve stem, or by galled threads. Tight packing glands can be loosened, and wheel wrenches can be used to provide mechanical advantage on stubborn valves. The cause of the binding should be identified and remedied. When reassembling valves, it is good practice to coat the threads with anti-seize compounds to prevent galling.

In replacing a flanged valve, the valve must be installed concentrically to the piping. Alignment may be made using spud wrenches to pry piping into alignment. The flange bolts may be inserted into every other hole and tightened evenly until enough friction develops to hold the valve and piping in proper alignment. The remaining bolts may then be inserted and tightened evenly in opposing pairs around the bolt circle.

If gaskets are to be replaced, the flanges should be completely cleaned of gasket residue. Cleaning is often done with sharp scrappers or fine emery cloth. It is common to see fine concentric patterns in the flange surfaces from the machining process. These machining tool marks are normal and help flange sealing. When cleaning the flange surfaces, care must be taken to avoid scratching radial marks that might form a path of future leakage.

Whenever a bonnet is removed for servicing of the valve disk or seat, it is good practice to completely remove the valve stem and clean all parts. Upon reassembly it is also good practice to apply an anti-seize compound to all threaded parts to avoid galling, ensure free operation of moving pieces, and future disassembly. Gaskets and packing must be replaced. Copper gaskets can usually be reused if in good condition, but should be annealed first. Soft gaskets should be coated with an appropriate sealing/anti-sticking compound and spiral-wound metal gaskets should be coated with an anti-seize compound.

21.5 Vent and sounding tubes

Tank vents allow air to be expelled when a tank is filled, and permit air to be drawn into the tank as it is emptied. The vents thereby prevent over-pressurization of the tank during filling, and the formation of a vacuum during pump out. Either of these conditions could result in buckling damage to the tank.

Sounding tubes are used to measure the level of fluid in a tank. The volume of fluid in the tank is then obtained from the ship's tank tables.

Vent pipes lead from the tank top up to the weather deck, terminating in a bell-like fitting in a gooseneck (Fig. 21.11). The gooseneck usually contains a ball float that acts as a check valve during heavy seas. For fuel-oil tanks, the vent contains a **flame-arresting screen.** This screen prevents a flame from working its way into the vent in the event of a nearby deck fire.

Sounding tubes extend through the tank top and terminate just above the bottom of the tank. At the bottom of the tank below the sounding tube is a **striker plate** to prevent damage to the inside skin of the ship as a sounding bob is lowered.

Access to the sounding tubes depends on the location. Sounding tubes that originate from a weather deck or interior passageway are usually fitted with tapered plugs, flush mounted into the deck. An alternative is a standing pipe rising about three feet (one meter) above the deck. These stand pipes are fitted with spring-loaded, quick-closing gate valves.

Tanks are sounded with tapes and bobs (Fig. 21.12). Sounding bobs may be of the plumb type or the ullage type. They are made of brass so that they are non-sparking for sounding fuel tanks. The **plumb bob** has a pointed end and is lowered on a measuring tape until it strikes the tank bottom. The tape is then reeled-up until the fluid mark is observed on the tape. If the tank contains water, it is common to apply chalk to the tape to better observe where the tape became submerged.

The **ullage bob** has a cupped end. As the bob is lowered down the sounding tube, it is jerked gently. When the ullage bob strikes the top of the liquid level, a popping sound will be heard up through the sounding tube. At that point, the measurement is taken from the top of the liquid to the top of the sounding tube, a distance called the **ullage**. After subtracting the ullage from the known depth from the sounding tube top to the tank bottom, the tank level is obtained. This method results in less wetting of the tape, a distinct advantage when sounding messy fuels.

CHAPTER 22
HEAT EXCHANGERS AND OTHER AUXILIARY MACHINERY

22.1 Heat exchangers

A heat exchanger is a device that transfers heat from one fluid to another. A **cooler** serves to remove heat from a warm or hot fluid, such as lubricating oil, by transferring the heat to the **coolant**, which is often seawater. A **heater** is used to add heat to a fluid, as in a feedwater heater, where the heating fluid is usually steam. There are many heat exchangers aboard ship. In size they range from very large units like the main condensers on steam ships, to small ones the size of a soda can. Two of most common types of heat exchangers are the **shell-and-tube type** and the **plate type**.

Fig. 22.1 shows a lubricating-oil cooler of the shell-and-tube type. In this heat exchanger, lubricating oil from the bearings of a steam turbine or a diesel engine is cooled by seawater. The seawater flows through the tubes and the oil flows around the outsides of the tubes. The seawater is the **tube-side** fluid, and the oil is the **shell-side** fluid.

In Fig. 22.1 seawater enters at the bottom left. A horizontal **baffle plate** directs the incoming seawater into the tubes in the lower half of the tube bundle. After it emerges from these tubes at the right end, the seawater makes a 180° turn and flows from right to left in the tubes in the upper half of the tube bundle. Thus, the seawater makes two **passes** through the exchanger. Oil enters through the oil connection at the top near the right end. It makes a single pass inside the **shell** and exits through the oil connection at the top near the left end. A **tube sheet** at each end keeps the oil and water separate.

Fig. 22.2 shows a typical plate-type heat exchanger. This heat exchanger consists of a stack of thin metal plates sandwiched between two covers. One fluid flows on one side of each plate, and the other fluid flows on the opposite side of each plate. Heat is transferred through the plate from one fluid to the other. Plate-type heat exchangers are more compact than shell-and-tube heat exchangers, but plate-type heat exchangers can only be used with low-pressure fluids.

Fig. 22.3 shows a single plate. There are four ports in each plate, one at each corner. In this example the ports on the left are associated with one fluid, and the ports on the right are associated with the other fluid. If the port is surrounded by a **gasket,** that fluid can not flow across the plate. If there is no gasket, the fluid can flow across the plate. In Fig. 22.3 there gaskets surround the ports on the right, but they do not surround the ports on the left. Therefore, fluid flows across the plate from the bottom left port to the top left port. On the other side of this plate the gasket positions are reversed so that the other fluid flows across the back of the plate.

In a steam cycle, a list of heat exchangers may include drain coolers, the gland-exhaust condenser, the first-stage feedwater heater, and high-pressure feedwater heaters, in addition to the condensers. For most diesel engines the combustion air is cooled between the turbocharger and the engine in a **charge air cooler**; the treated freshwater that cools the cylinder liners and cylinder heads of diesel engines is cooled by seawater in a **jacket water cooler** and sent back to the engine again. Some large diesel engines have water-cooled pistons and **piston water coolers**. An air compressor may have a heat exchanger between the stages and another one after the final stage. These are called **intercoolers** and **aftercoolers** respectively. Other examples of heat exchangers aboard ship include lubricating-oil coolers and, for ships using heavy fuels, fuel-oil heaters.

22.2 Condensers

A condenser is a shell-and-tube heat exchanger with a specialized function – the condensation of steam to water. Condensers serving steam turbines generally operate at pressures well below atmospheric on the steam side, because this enables the associated turbine to produce more power by exhausting at a lower pressure.

In a condenser heat is transferred from the steam to cooling water, usually seawater. As heat is removed from the steam, it condenses to liquid water called **condensate**. This condensate is then pumped back to the boiler as feedwater. The seawater is drawn in from the sea, passed through the tubes in the condenser, and discharged back into the sea at a slightly higher temperature. When steam changes to water, its volume decreases drastically. This reduction in volume is responsible for creating the vacuum in the condenser, although air ejectors are necessary to maintain the vacuum.

The propulsion turbines of a steam ship exhaust to a very large **main condenser**. Fig. 22.4 shows a typical main condenser for a naval ship. If turbines are used to drive electric generators, there is normally a separate condenser, called an **auxiliary condenser**, for the generator turbines. The auxiliary condenser is similar to the main condenser, but it is much smaller.

There are thousands of tubes in a typical main condenser. These tubes are supported by **tube support plates**, which also serve to direct the flow of steam on the shell side of the condenser. Steam leaving the low-pressure turbine enters the condenser at the top and flows down over the outsides of the tubes as it transfers heat through the tube walls into the seawater. This heat transfer causes the steam to condense to water on the tube surfaces. From there the water falls to the bottom of the condenser, where it is drawn out through the condensate outlet connections to the **condensate pumps**.

Seawater for the main condenser enters a large inlet pipe through an opening in the ship's hull. In Fig. 22.4 seawater enters the condenser through the large connections at the far right. Note that in this case there are two inlet openings. One comes from the **main circulating pump**, which pumps seawater through the condenser. The other opening is for **scoop injection**. When ships operate at sufficiently high speed, the motion of the ship through the water is sufficient to force water through this connection, and no pump is required. Leaving the condenser, seawater returns to the sea through another large pipe, the **overboard discharge pipe**. The temperature of the seawater may rise by about 10°F as it flows through the condenser.

Air is removed by an **air ejector** or a **vacuum pump** that is connected to the air off take connection, shown to the left of the steam inlet in Fig. 22.4. Air ejectors are described in section 20.6, and vacuum pumps are described in section 20.7.

22.3 Freshwater generators

Oceangoing ships are fitted with freshwater generating facilities to produce freshwater from seawater. The freshwater is used for washing, cooking, and drinking (**potable water**) and for use as make-up feed to compensate for losses from steam systems and freshwater cooling systems. Most freshwater generators fitted aboard ships are **evaporators** (also called **distillers**), which utilize the fact that when seawater is boiled the vapor is pure: when the vapor is collected and condensed the resulting **distillate** is freshwater.

Almost all modern evaporators are of the low-pressure type, meaning that they operate at pressures below atmospheric, i.e. under a vacuum. Two facts justify this complication. First, at low pressure, the seawater will boil at a low temperature, in turn enabling the use of a low-temperature heating source like the jacket cooling water from a diesel engine or low-pressure bleed steam from a steam turbine. Second, when seawater is boiled at or above atmospheric pressure it leaves a hard deposit (called **scale**) on the heating surface, which interferes with heat transfer and is hard to remove. At the lower boiling temperatures associated with pressures below atmospheric this scaling is much reduced.

Fig. 22.5 shows a low-pressure evaporator of the type often used aboard motor ships. Seawater is boiled in the evaporator shell using heat from the engine cooling water. The resulting vapor, which is free of salts, passes to the distillate condenser, where it condenses to form the pure distillate. The distillate pump transfers the distillate to the potable water tanks or make-up feed tanks. Note that the seawater pump circulates seawater through the tubes of the distillate condenser, and then, through a branch line, also supplies seawater feed to the evaporator shell. A second pump, the eductor pump, provides seawater at a higher pressure to drive the air eductor, which maintains the vacuum in the evaporator shell and the distillate condenser. The eductor pump also drives a brine eductor, which removes seawater from the evaporator shell to prevent the salt concentration of the seawater in the shell, called **brine**, from becoming excessive.

Fig. 22.6 shows a double-effect, flash evaporator often fitted to steam ships. It is a **flash evaporator** because the seawater feed, which is heated under pressure in the feed heater by low-pressure steam, does not begin to vaporize until it enters the first effect, which is maintained under a vacuum. Since the seawater does not boil in the heater, scale will not form on the heating tubes. Only in the first effect does some of the water "flash" to steam.

The evaporator of Fig. 22.6 is called a **double-effect evaporator** because evaporation occurs in two successive steps: after some of the seawater feed has flashed to vapor in the first effect, the brine flows to the second effect, where, since the second effect is maintained at an even lower pressure, more vapor is produced. As a result of this two-stage evaporation the amount of distillate produced per unit of energy used to heat the feed is higher than in single-effect evaporators. Although this increased efficiency is not important on most motor ships, where there is sufficient waste heat available in the engine cooling water, it is important on steam ships where live steam is used as the heating source for the evaporator.

These figures are simplified to illustrate the principles. In reality an evaporator may appear very complex, with the components packaged into a compact unit, and with many connections, pipes, valves, pumps, eductors, and gages for seawater, brine, jacket water, steam, vapor, distillate, and vacuum. An example is shown as Fig. 22.7.

Where neither steam nor engine-cooling water is available as a heating source, a **vapor-compression distilling plant** or **reverse osmosis plant** might be fitted.

22.4 Purifiers

The large quantities of lubricating oil used in main propulsion engines, turbines, and reduction gear sets must be cleaned or purified. This process involves removing any water or solid particles from the oil and then returning the oil to service. On motorships that burn heavy fuel, the fuel oil must be purified before it is injected into the engine. **Purifiers**, also called **separators** or **centrifuges**, are provided for these services. Although the details differ, the basic design of fuel and lubricating-oil purifiers is similar.

Figs. 22.8 and 22.9 show a typical purifier. With reference to the section view, the heavy bowl and the stack of conical disks that it contains are rotated by the spindle, which is driven at high speed through gearing by the electric motor visible in the photograph. Untreated oil flows in at point 1 and is directed to the bottom of the bowl. Holes in the disks allow the oil to flow upward through the stack. The rapid rotation of the disks and bowl spins the mass of oil. Since the water and sludge are heavier than the oil, they are forced outward by the centrifugal force, while the oil remains nearer the center. Sludge accumulates on the inside periphery of the bowl, and water, just inside of the sludge. The water passes upward outside the outer edge of the disk stack and exits at point 5. The top disk directs clean oil toward the center, and the clean oil exits at point 4.

Like most modern oil purifiers, the unit in these figures is said to be self-cleaning or sludge-ejecting: the bottom of the bowl can move down slightly, but is normally forced upward against the upper part of the bowl by water introduced under pressure at point 7. At timed intervals while the purifier is running, this water pressure is interrupted momentarily, and accumulated sludge is forced out by centrifugal force through ports in the bowl at point 3. Purifiers that are not self-cleaning must be disassembled and cleaned by hand frequently. Even self-cleaning purifiers must occasionally be cleaned manually.

22.5 Sewage treatment plants

Wastewater from toilets and urinals is classified as **black water**. Wastewater from showers, sinks, washbasins, and laundry facilities is classified as **gray water**. In general, black and gray water can be discharged directly overboard at sea, but in port and in coastal and inland waters, black water must be treated before it is discharged. However, in some areas gray water must be treated, and there are zones where no discharges of even treated wastewater are allowed.

To meet the requirements, wastewater is processed through a sewage treatment plant, or MSD (Marine Sanitation Device). An example is shown in Fig. 22.10. Most MSDs aboard ship are self-contained, packaged units that utilize bacterial action. Certain types of bacteria, in the presence of oxygen dissolved in the water, act upon waste products to produce carbon dioxide, water, and harmless sludge. The wastewater is strained and collected in tanks that are equipped with **aerators**. These devices inject air into the water to encourage the bacterial action. Once bacterial action is complete, the water is usually chlorinated or exposed to ultraviolet radiation to sanitize it. The sanitized products are then ready for discharge as permitted. The sludge may be discharged overboard at sea or, in some cases, notably passenger ships, the sludge is dried and burned in an incinerator.

22.6 Oily-water separators

Oily-water separators are used to clean bilge water so that it may be discharged overboard. In this service the rate of water flow through the separator is low, and the oil is present in only trace quantities. The principle of coalescence is used: at very low flow rates the small droplets of oil tend to concentrate into larger globules, which are more easily separated from the water by gravity. Oily-water separators are static, self-contained units that operate automatically. A typical unit comprises a number of cylindrical tanks, or a single cylindrical or rectangular tank divided internally into chambers. A feed pump draws from an oily-water collecting tank or bilge holding tank, usually in the engine room double bottom, and discharges the oily water through a strainer into the first tank or chamber, where it is usually heated. The water then flows at low velocity over a large number of closely spaced, sloping steel plates, which encourage the process of coalescence and provide a path for the larger droplets to rise towards the surface, where the accumulated oil can be stripped off. The water can be passed to another chamber where the process can be repeated. With a sufficient number of stages, usually two or three, the water is sufficiently clean to discharge overboard. The collected oil can be passed to a heavy fuel-oil settling tank or to a waste-oil tank.

CHAPTER 23
PIPING SYSTEMS

Piping, fittings, valves, pumps, heat exchangers, tanks, and other machinery and equipment are joined together to form systems. These systems, although often interrelated, are dedicated to performing specific functions on board ship. Systems are identified by the fluid they convey and may be further identified by their function. Some of the major propulsion-related systems are the fuel-oil service, fuel-oil transfer, main steam, auxiliary steam, condensate, feedwater, seawater circulating, seawater service, and lubricating-oil service, ship's-service compressed air, control compressed air, and lubricating-oil transfer systems. Other systems such as heating, air conditioning, potable water, sanitary water, and sewage are necessary for crew support. In this chapter, simplified flow diagrams and brief descriptions of most of the major shipboard systems are presented. For clarity, some redundant (standby) equipment may not be shown. These diagrams show typical systems, while systems found on individual ships are likely to be somewhat different. Refer to Fig. 23.1 for a list of symbols referenced in the subsequent flow diagrams.

23.1 Fuel-oil transfer system

The fuel-oil transfer system (see Fig. 23.2) is used to transfer fuel from the ship's fuel-oil storage tanks, or **bunkers**, to the ship's **settling tanks**. At sea, the fuel-oil transfer system is used for replacing the daily consumption of fuel from the settling tanks (also called **settlers**). Although the diagram shows only double-bottom storage tanks, many ships store fuel in wing tanks or deep tanks. Occasionally, the fuel-oil transfer system may be used to transfer fuel from one storage tank to another for adjusting the ship's trim and stability. The fuel-filling system used for **bunkering** the ship is tied into the fuel-oil transfer system.

Fuel-oil transfer pumps are usually vertical, rotary screw, positive-displacement pumps, capable of taking suction from any tank and pumping fuel to any other tank through the system and the **manifolds**. A manifold is a series of valves that are grouped together in a single housing, consolidating operation to a single location. In the fuel-oil transfer system, each manifold is divided, with a suction valve for each tank on one side, and a discharge (or fill) valve, on the other. A single pipe line connects each tank to its pair of manifold valves. The suction valves are connected to the transfer-pump suction for removing fuel from the tank, and the discharge valves are connected to the transfer-pump discharge and to deck-fill stations for filling the tanks. A second fuel-oil transfer pump may be fitted as a standby unit, or a different fuel pump may be provided with cross-connecting piping to serve the purpose

Vents are fitted to all fuel tanks, led to the weather deck to allow air to enter or leave when the pump is emptied or filled. Inside heavy fuel-oil tanks are heating coils that are used to warm the fuel to reduce its viscosity prior to transfer, to facilitate pumping. Generally, steam is circulated through the heating coils, but on some ships a non-combustible **thermal fluid** is used. Transfer piping may be fitted with heat-tracing systems to reduce fuel viscosity in the piping. A simplex suction strainer is provided for the fuel-oil transfer pump. Because they are positive-displacement pumps, each transfer pump is fitted with a relief valve from the discharge back to the suction to protect the pump and system against over-pressurization. The pump is also fitted with a manual recirculation line from its discharge back to its suction to reduce pumping rates if cold oil must be pumped, which could cause high motor loads.

23.2 Fuel-oil service systems

The fuel-oil service system is used to take the daily supply of fuel from the settling tanks and deliver it at the proper temperature, pressure, and viscosity to the fuel burning equipment used for propulsion power. While similar in intent, the fuel-oil service systems of steam and diesel vessels vary in configuration and are covered separately.

23.2.1 Fuel-oil service system - steam ships

The fuel-oil service system (see Fig. 23.3) is used to take fuel from the settling tanks and supply it to the burners at the boilers under sufficient pressure and temperature so that the fuel can be properly atomized for complete combustion.

The fuel-oil service system takes the fuel from a settler via a high or low suction. Normal operation is to draw the fuel from the low suction. In the event that water (from seawater intrusion or leaking heating coils) is encountered, the engineer can shift to the high suction. The water is eliminated later, often by **cracking** (i.e., by opening slightly) the low suction, over the next week or so until the water is "burned-off," at which time the low suction is placed back in service. Alternatively, the water can be drained to an oily-bilge tank. Although only one settler is shown for clarity, two are usually fitted. Sometimes four settlers are provided and the high suctions are eliminated.

Oil from the settlers passes first through a coarse-mesh **suction strainer**, then through a fuel-oil service pump (two are usually fitted; only one is shown), then through fuel-oil heaters, to a fine-mesh **discharge** (or **high-pressure**) **strainer**. The discharge strainers are designed to filter out scale and other foreign matter that might otherwise damage or clog the burner nozzles and impair combustion. Fuel-oil service strainers are always of the **duplex** type so that the strainers may be changed and cleaned without taking the system out of service. Each duplex strainer is fitted with a differential-pressure gage to indicate when the strainer is fouled. Normal practice with heavy oil is to change and clean the strainers daily.

Heating of heavy fuel oil begins in the storage tank where it may be heated to about 100°F (40°C) to facilitate transfer. The fuel is further heated in the settling tanks. For safety, the settler temperature is limited to about 120°F (about 50°C), safely below the **flash point** of the fuel. (The flash point is the temperature at which the fuel vapors would ignite in the presence of a spark.) In order to properly atomize and burn the oil, the fuel is then heated in the fuel-oil heaters to about 200°F (90°C) to reduce the viscosity sufficiently for good atomization. Multiple heaters are used for redundancy and to adjust the fuel heating to match requirements.

The system in Fig. 23.3 shows only a single boiler. In actuality, the system branches after the fuel meter to supply each boiler, with the equipment shown duplicated in each branch.

The fuel-oil service system also includes a manual **recirculation valve** at each burner manifold, piped back to the pump suction or to the settlers. This valve is cracked before lighting off a cold boiler to bring hot fuel to the burners for good atomization at light off. The recirculation valve is also used to purge the lines back to the settlers in the event of water contamination.

23.2.2 Fuel-oil service and treatment system - motorships

Most diesel-driven merchant ships use heavy fuel oil (**HFO**) in the main engine most of the time, and only occasionally use distillate fuel oil (**DO**) for propulsion. The fuel-oil system (see Fig. 23.4) must be capable of handling the variations in fuel viscosity, and changing between the fuels. The HFO system includes one or two settling tanks, but to further clean the fuel, it is passed through **fuel-oil purifiers** to a **day tank** before it enters the fuel-oil service system.

Heavy fuel must be heated to reduce its viscosity, but temperatures in the settling tank and day tank are kept below the flash point. Prior to entering the purifiers the fuel is heated further to facilitate separation of water and sediment, but here the temperature is limited by the boiling point of water. Normally two or more **heavy-fuel purifiers** are provided, arranged for simultaneous operation to maximize water and sediment removal.

Fuel flows from the day tank to the **mixing tank** (often a large pipe), from which the **booster pumps** discharge it through the **final heaters** to the injection pumps on the engine. At the final heaters, the fuel is heated to a temperature that will ensure a viscosity low enough for atomization and combustion. This temperature can be as high as 300°F (150°C). A **viscosimeter** controls the steam supply to the final heaters.

The power output of the engine is controlled by the amount of fuel discharged by the injection pumps, and excess fuel is recirculated back to the mixing tank.

The distillate fuel system parallels the heavy-fuel system, with a settler, a purifier, and a day tank, but distillate fuel does not normally require heating. In the example shown, the distillate fuel is not only available for occasional use in the main engine, but also for full-time supply to the ship's-service diesel generators and the auxiliary boiler. In many other cases the generators are normally run on heavy fuel.

23.3 Steam, condensate, and feedwater systems - steam ships

The basic steam cycle (see Fig. 23.5) can be divided into three fundamental piping systems: steam, condensate, and feedwater, each of which is discussed separately in the following sections.

23.3.1 Main steam system

Main steam is high-energy, superheated steam used for main propulsion power and electricity generation; **auxiliary steam** is used for other services.

See Fig. 23.6. Main steam comes directly from the superheater outlets of the boilers. At each boiler, the superheated steam line branches into three lines, one going to the main propulsion turbines, another going to the turbogenerators, and the third going to the desuperheater to supply auxiliary steam. **Main-steam stop-check valves** are fitted at each superheater outlet for the main propulsion turbines and for the turbogenerators. These valves prevent reverse flow of steam to a boiler that has tripped off line. Each of these lines contains a second valve, customarily referred to as the **bulkhead valve**, as it was located on the engine-room side of a watertight bulkhead on older ships with the engine room separated from the fireroom. The main steam line supplying the **desuperheater** is not fitted with a valve.

From the bulkhead valves, main steam to the propulsion turbines passes to **ahead and astern throttle valves**. The throttle valves are used to vary the amount of steam to the turbine to control the ship speed and direction. In the line between the astern throttle and the astern turbine is the **astern guardian valve**. Under normal steaming at sea the astern guardian valve is closed tightly. The astern guardian valve is opened only during maneuvering, or during periods of low visibility or heavy traffic. Many variations to this typical piping arrangement can be found aboard ship.

23.3.2 Auxiliary steam system

Auxiliary steam is used for auxiliary equipment and for heating. See Fig. 23.7. **Desuperheated steam** comes from the desuperheater. All of the steam generated is first superheated, and auxiliary steam is branched from the superheater outlet to the desuperheater. The typical desuperheater is of the **submerged-tube** type, where the steam is passed through a tubular heat exchanger located under the water level in the steam drum, or in the water drum. (In some applications attemporating-type desuperheaters are used, where feedwater is sprayed directly into the steam to cool it.) Auxiliary steam supplied from the boilers via the desuperheaters is **live steam**.

Live steam at full boiler pressure is commonly used to drive the main feed pumps and for sootblowing. Other services, which may include other steam-driven pumps, steam air ejectors, distiller-heating steam, fuel-oil heaters, tank-heating coils, and domestic hot-water heaters require steam at lower pressure, and at sea speed some of this equipment may be supplied by turbine-bleed steam. Feedwater heaters are supplied only with bleed steam. Where auxiliaries have different steam-pressure requirements, they are supplied via a series of pressure-reducing valves in an arrangement that cascades higher-pressure systems to lower-pressure systems.

23.3.3 Auxiliary exhaust steam system

The **auxiliary exhaust system**, also called the **back-pressure system** and the **maintained exhaust line**, is supplied from the steam exhausted from the auxiliaries, principally the main feed pump. See Fig. 23.8. The thermal energy of the auxiliary exhaust steam is recovered by using this steam as the heating source for the DC heater, and in some installations the boiler air heaters. Maintenance of constant pressure in this system is very important, and feed pump and DC heater problems can often be attributed to problems with the back-pressure system. The exhaust from the feed pump is usually supplemented with make-up steam, using bleed steam when available at sea speeds, otherwise with live steam. When the back pressure is too high, the excess steam is automatically dumped to either the main or auxiliary condenser.

23.3.4 Condensate system

Refer to see Fig. 23.9. Condensate collects in the bottom of the condenser in the drainwell or **hotwell**. The condensate pump takes suction from the hotwell and discharges through several heat exchangers to the **DC heater**, which is described below. Beyond the DC heater the condensate is called **feedwater**.

The first heat exchanger encountered by the condensate may be the **air-ejector condenser**. This air-ejector condenser is divided internally into two sections, the **intercondenser** and the **aftercondenser**. A steam-jet air ejector for a main condenser has two stages. The first-stage air ejector draws air and other **non-condensable gases** from the main condenser and exhausts to the intercondenser. Water from steam condensed in the intercondenser returns through a **loop seal** to the condenser. The second-stage air ejector draws the gases from the intercondenser and discharges them to the aftercondenser. Water from steam condensed in the aftercondenser is generally returned to the **atmospheric drain tank (ADT)**, sometimes called the **freshwater drain-collecting tank (FWDCT)**. Thermal energy rejected by condensing the air-ejector steam is recovered by heating the condensate.

The next heat exchanger in the circuit is likely to be the **gland-exhaust condenser**, also known as the gland leak-off condenser. This heat exchanger recovers heat and vapor by condensing leak off steam from the main turbine glands drawn in by the **gland exhaust fan**. It may also receive vapor and non-condensable gases from aftercondenser and DC heater vents. The resulting drains are generally returned to the ADT.

The condensate pumps are usually vertical, two-stage centrifugal pumps, designed with a low suction head requirement, which means that they are able to extract the condensate despite the vacuum in the condenser. The condensate system is fitted with a **recirculation line** whose purpose is to ensure that there is adequate flow through the air-ejector and gland-exhaust condensers and that the pump does not run dry. Recirculation is automatically controlled by a thermostatic valve measuring the air-ejector outlet temperature, and/or by hotwell water level. A manual recirculation valve is often cracked during maneuvering as a matter of general practice to ensure adequate condensate flow at low power.

The next heat exchanger may be a **first-stage heater** or **low-pressure heater** which uses bleed steam from a low-pressure turbine extraction to heat the condensate. It is common practice to combine the gland-exhaust condenser and the first-stage heater into a single unit. Another heater may follow, but ultimately the condensate is directed to the DC heater.

Condensate from clean low-pressure drains such as space heaters and hot-water heaters is returned directly to the ADT. The ADT contents are returned to the system either via a float-controlled, vacuum-drag line to the main or auxiliary condensers, or a float-actuated ADT pump that discharges the hot water into the condensate line prior to the DC heater, as in the example shown.

Condensate from potentially oily low-pressure drains such as the fuel tank heating coils or fuel-oil heaters is returned to a **contaminated-drain inspection tank (CDT)**. The CDT is fitted with weirs and baffles to trap floating oil from returning easily to condensate and feedwater systems. The CDT is fitted with back-lit viewing ports or oil detectors so that the condensate can be observed to be oil-free. Under normal operation the CDT overflows into the ADT where the condensate is returned to the system. However, if oil is observed in the returns, they must be diverted to an oily bilge until the source of contamination is found and eliminated.

23.3.5 Feedwater system

The feedwater system (see Fig. 23.10) is considered to start at the **direct-contact heater**, known also as the **DC heater**, the **deaerator**, the **DA tank**, the **deaerating feed tank**, or the **DFT**, a large tank that, on merchant ships, is generally located high in the engine room. The condensate is called **feedwater** after entering the deaerator.

The DC heater has several functions. First it serves to **deaerate** the feedwater before it enters the boiler, where, at high temperature, the oxygen would cause corrosion. Deaeration is accomplished by heating the feedwater to the boiling temperature corresponding to the pressure in the DC heater, called the **saturation** temperature. At the saturation condition, dissolved gases are insoluble in water and tend to separate. This heating is accomplished by mixing the condensate directly with auxiliary exhaust steam. To aid the deaeration process, the condensate is sprayed into the DC heater in a direction that provides mechanical agitation and thorough mixing with the steam flow.

Non-condensable gases vent from the DC heater through a heat exchanger called the **vent condenser**. The vent condenser uses incoming condensate as the coolant to reduce the quantity of water vapor that is vented along with the non-condensable gases. The gases and some water vapor are vented to the gland exhaust condenser as noted above, where the water vapor is recovered and the non-condensable gases are vented to the atmosphere via the gland exhaust fan.

The second function of the DC heater is act as a **surge tank.** The DC heater stores feedwater for the boilers for normal transient conditions. It also provides several minutes of boiler feedwater in the event of a plant casualty: generally the DC heater is sized with enough capacity to supply the boilers for ten minutes at full power.

The third function of the DC heater is feedwater heating to improve cycle efficiency.

The location of the DC heater high in the engine room serves to raise the pressure on the inlet to the main feed pump. The resulting **net positive suction head (NPSH)** at the feed pump inlet, which is the pressure above the feedwater vapor pressure, is required to keep the hot feedwater from flashing to steam as it enters the pump impeller. If flashing occurs, the feed pump will overspeed and trip.

An alternative to locating the DC heater at a high elevation is to use a **feed booster pump** in series before the main feed pump. The feed booster pump is a low-head pump that can meet the suction requirements of the main feed pump but does not have severe suction requirements itself. Feed booster pumps are commonly found on naval vessels in order to consolidate all propulsion machinery into a small volume.

The feed pump raises the feedwater pressure sufficiently above boiler pressure with enough allowance to overcome pressure losses in piping and feedwater regulators. Typically, the main feed pump requires several hundred horsepower and is usually driven by a non-condensing steam turbine. The exhaust from the feed-pump turbine is the prime steam source to the DC heater via the auxiliary exhaust system. The turbine speed is governor-controlled to match the pump output to the steaming requirements.

To provide an immediate alternative supply of feedwater to the boilers, the feed-pump discharge is split into two lines at the pump called the **main feed line** and the **auxiliary feed line**. The main feed line is the normal feedwater supply to the boilers and in the branch to each boiler, contains a stop-check valve, an automatic feedwater regulator, and stop valves. The **automatic feedwater regulator** controls the flow of feed to the boiler to maintain the water level in the steam drum. In each auxiliary feed line the stop valves are normally open, and only the auxiliary stop-check valve is closed. All of the stop-check valves are operable from the firing platform through reach rods, to facilitate manual control of feedwater flow.

An automatic feedwater regulator controls the flow of feedwater to the boiler primarily in response to boiler water level. A **two-element** feedwater regulator also uses steam flow rate to provide better level control by reducing the effects of **shrink and swell** (see Chapter 4). In a three-element regulator, the control system also responds to feedwater flow rate.

To accommodate low feedwater flow rates when maneuvering or in port, the main feed pump is fitted with a recirculation line from the pump discharge back to the DC heater. This recirculation line ensures that there is a continuous flow of feedwater through the pump even during shut-off conditions, when the boiler level is high and the feedwater regulators are closed. Without a continuous flow through the pump, the water would overheat and flash to steam, and the pump would overspeed and trip.

It is common to find the feed pump fitted with trip mechanisms for low lubricating-oil pressure, overspeed, high back pressure (high exhaust-steam pressure) and hand tripping. It is necessary to manually reset the governor after the pump trips.

On most ships, the feedwater flows from the feedwater regulator to the steam drum via an economizer, a heat exchanger in the uptake of each boiler. The economizer recovers heat from the boiler flue gases to heat the feedwater, thereby improving efficiency. However on many ships a **rotary-regenerative air heater** is fitted in lieu of the economizer (see Chapter 4). These ships are fitted with one or two high-pressure feedwater heaters after the DC heater and feed pump. The high-pressure heaters use bleed steam extracted from the high-pressure turbine.

23.3.6 Make-up feed system

The steam plant operates on a closed cycle, but some normal losses and leakages occur which must be replaced with **make-up feed** supplied from the distilled-water tanks (also called reserve-feed tanks). The distilled-water tanks are filled with extremely pure water from the ship's evaporators. The level in the DC heater, which allows for surges of water in the system, controls make-up feed intake. See Fig. 23.11. If the level in the DC heater drops enough, a controller sensing the DC heater level automatically opens the make-up feed valve. Make-up feed may be introduced into the system via a vacuum-drag line from the distilled-water tank to one of the condensers as shown, or at the atmospheric drain tank. When the DC heater level is normal, the make-up valve closes.

If the DC heater level becomes too high, another controller opens a condensate **dump valve**. This dump valve branches from the condensate pump discharge line after the gland-exhaust condenser, diverting condensate to the distilled water tank. The dumped condensate has been utilized to cool the air-ejector and gland-exhaust condensers, while hotter DC-heater water is kept in the system for economy.

23.4 Waste-heat recovery steam systems - motorships

Motorships require steam for general heating services and for heating heavy fuel oil. Most large motorships recover heat from the main-engine exhaust to generate steam in a **waste-heat boiler (WHB)**. Since the main propulsion engine is secured when the ship is not making way, an oil-fired auxiliary boiler is also provided. Typical systems are shown in Fig. 23.12.

The primary sources of waste heat from diesel engines are the hot exhaust gases and the jacket water. The exhaust gases provide a high-energy source capable of producing reasonably high steam pressures that may be used for turbine-driven equipment. Jacket water is used directly as the heat source for the ship's evaporators.

Some smaller motorships recover waste heat by heating a non-combustible fluid, which is circulated to equipment requiring heating. This system is referred to as a **thermal-fluid system**.

Figs. 23.12A and B show two of many arrangements of steam systems found on modern merchant ships. To ensure that there is always an adequate supply of steam to the ship, regardless of the status of the diesel plant, an oil-fired auxiliary boiler is installed. This boiler is automated to light-off and generate steam in the event that steam pressure drops. In these examples, the oil-fired boiler provides the separating space for the steam generated in the waste-heat boiler. Often, the waste-heat boiler has its own steam drum.

Fig. 23.12A shows a basic waste-heat recovery system that generates saturated steam used primarily for heating fuel oil. The boiler pressure is typically 100 psig (7 bars) or less. General heating steam is obtained through pressure-reducing valves.

Fig. 23.12B shows a steam plant with a turbogenerator. This system is practical only on high-powered ships where there is sufficient exhaust gas to generate enough steam to, in turn, generate a significant amount of electricity. The waste-heat boiler uses an economizer section in the coolest area of the uptakes to preheat feedwater before entering the boiler. Saturated steam from the boiler is sent to heating services, or is passed through the superheater to the turbogenerator.

At high main engine power levels more exhaust gas and waste heat is available than can be used. In these examples, the excess steam generated is dumped to a condenser via a back-pressure regulating valve. Another approach uses an exhaust-gas bypass duct in the uptake, with dampers to direct some of the exhaust gas flow away from the waste-heat boiler.

23.5 Steam systems for tanker cargo services

Most tankers require steam for cargo services, which may include cargo heating at sea and cargo pumping in port. Hot water for hot tank cleaning may also be required. The quantity of steam can be very substantial.

When steam is used for cargo heating, there is a possibility that the condensate can become contaminated with oil from leaks in the tank heating coils. This oily condensate must be kept from entering oil-fired boilers. A common arrangement fitted on many tankers segregates the cargo-heating system by using a large, shell-and-tube heat exchanger called a **contaminated evaporator** or **low-pressure steam generator (LPSG)** (see Fig. 23.13). Live steam or turbine-bleed steam circulates through the tubes of the LPSG, boiling water on the shell-side of the tubes. This live steam or bleed steam is condensed and returned to the boilers via the atmospheric drain tank. The secondary steam produced on the shell side of the LPSG is used for cargo heating, and its condensate is returned to the LPSG via a **contaminated-drain inspection tank (CDT)** and LPSG feed pumps.

Cargo- and ballast-pump turbines on steam ships are usually supplied with desuperheated steam and may exhaust to the main condenser or to a separate cargo-pump condenser. On diesel-powered tankers, the cargo- and ballast-pump turbines are usually supplied directly from oil-fired cargo-service boilers, which generate saturated steam, and the turbines exhaust to an independent cargo-pump condenser. On these ships, the condensate from the cargo-pump condenser is the feedwater for the cargo-service boilers in a simple closed cycle.

23.6 Seawater circulating and cooling water systems

23.6.1 Seawater circulating and cooling water systems - steam ships

Seawater systems can be classified as low-pressure seawater circulating systems, or as higher-pressure seawater cooling (seawater service) systems. Circulating systems (see Fig. 23.14) operate at pressures generally below 30 psia (2 bars). High-volume, low-head centrifugal pumps, often of the propeller or mixed-flow type, are used. These circulating pumps supply the quantities of water required to absorb the large amount of latent heat from steam exhausted to the condensers. Main circulators serve the main condenser, while a separate auxiliary circulator is usually provided for each turbogenerator auxiliary condenser. Circulator crossover connections provide redundancy.

Many ships have **scoop injection** for the main condenser, as in Fig. 23.14, where a piping branch is oriented from the ship's bottom to the condenser inlet so that the ship's forward motion "scoops" circulating water into the condenser. The scoop injection valve is opened after sea speed is reached, allowing the main circulating pump to be secured, reducing electric power requirements. The main circulating pump is smaller for these ships, adequate for reduced-power steaming and maneuvering. A check valve is installed in the scoop piping so that the main circulating pump can be started before the scoop-injection valve is closed.

The circulating-water pumps take suction from the **sea chests**. The sea chests are fitted with steam lines to melt ice that can clog the sea chest in frigid waters, and lines for flushing with high-pressure firemain water.

Typically **high and low sea suctions** are fitted. The high suction is used in port and during inland maneuvering to avoid sucking silt and debris from a harbor bottom. The low suction is used at sea to avoid sucking air into the circulating system during heavy rolling.

Because the circulating pumps are limited to low discharge heads, it is common to find open vent lines from the sea chests, terminating in the upper engine room, at an elevation above shut-off head rating of the pump. These open vents ensure that air is continuously vented and that the upper condenser tubes are always flooded with seawater.

The main lubricating-oil coolers may be supplied by a branch connection from the head of the main condenser. The flow is usually hand throttled using a butterfly valve on the lubricating-oil cooler outlet to maintain the desired oil temperature, usually around 110°F (about 45°C). The turbogenerator air and lubricating-oil coolers may be cooled either from auxiliary circulating water or from the higher-pressure seawater service system.

The higher-pressure cooling water system is the **seawater service** system. It generally operates above about 50 psia (3 bars). The seawater service system supplies auxiliary coolers such as the feed-pump lubricating-oil coolers, the refrigeration condensers, and the air-conditioning condensers, as well as stern tube flushing for water-lubricated stern tubes. Occasionally this system supplies the main lubricating-oil coolers as well.

23.6.2 Seawater circulating and cooling water systems - motorships

Seawater pumps on motorships are usually arranged to draw suction from a common seawater suction main which crosses athwartships from a high sea-suction chest on one side to a low sea-suction chest on the other side. Typically, a low-discharge-head, main-engine seawater pump, with an identical standby pump, draws from the main and discharges through the main-engine charge-air cooler, jacket-water cooler, lubricating-oil cooler, and where fitted, piston-water cooler and injector-cooling tank, and then overboard. Common practice is to install a three-way valve to partially recirculate some of the warmed circulating water back to the pump suction main to ensure that the cooling water is not too cold for the systems regardless of the seawater temperature. Usually a second pair of pumps, the auxiliary seawater cooling pumps, draw seawater from the common main for cooling of the ship's-service diesel generators and other auxiliary machinery.

Seawater service systems are prone to fouling and corrosion, and on many modern ships, much of the auxiliary machinery is cooled with freshwater from a central freshwater cooling circuit. The freshwater is, in turn, cooled in a central freshwater-to-seawater cooler. The seawater side of the central cooler is circulated by the auxiliary seawater cooling pumps. These low-pressure coolers are often of the plate type because of their compact size.

23.7 Freshwater cooling systems - motorships

Treated freshwater is used for internal cooling of diesel engines, thus avoiding corrosion problems associated with seawater. Auxiliary diesel engines on board ship generally have self-contained cooling systems using an engine-driven pump to circulate freshwater through the jacket for cylinder cooling, through the lubricating-oil cooler, and through the charge-air cooler. Emergency-generator engines use an engine-driven, fan-cooled radiator ducting air from the weather for rejecting heat. Auxiliary diesel-generator engines usually use a seawater-to-freshwater heat exchanger.

For propulsion engines, the freshwater systems may be divided into subsystems for charge-air cooling, lubricating-oil cooling, jacket-water cooling, and sometimes, for fuel-injector cooling and piston cooling, as shown in Fig. 23.15.

The jacket-cooling water has enough useful waste heat to be used as a heat source for the ship's distiller. An **expansion tank** is installed in a branch connection to allow for expansion of the cooling water as it heats. An air separator is installed in the line to ensure that the engine jacket does not become air-bound. Water losses are manually made-up to the expansion tank from the freshwater system.

If pistons are water-cooled, as in some large, slow-speed, crosshead diesels, the cooling water reaches the piston through telescoping tubes within the crankcase. Most engines use oil to cool the pistons, making a separate freshwater piston-cooling circuit unnecessary.

Injector-cooling water, if used, may branch from the jacket-water system, or may be a separate system as illustrated in Fig. 23.15.

23.8 Main lubricating-oil system - steam ships

The main lubricating-oil system is used to lubricate and cool the bearings of the main propulsion turbine and reduction gear, the gear meshes, and the main thrust bearing. It also provides operating oil for the turbine control and overspeed protection equipment. The main lubricating system includes a **purifier** that centrifuges the lubricating oil on a continuous basis to remove solid particles and water.

Two types of lubricating systems found on board ship are the **gravity system** and the **pressure system**. The gravity system is the most common in merchant service, while the pressure system, which is more compact, is more common on naval vessels.

In the gravity system, illustrated in Fig. 23.16, oil is collected in the main engine sump located in the double bottom beneath the reduction gear. The oil is drawn from the sump by a rotary pump through a coarse suction strainer. The strainer contains magnets that protrude into the strainer basket to pick out any ferrous chips that may contaminate the oil, thereby providing an early warning of a pending gear or bearing failure. The oil is then pumped through a discharge strainer of finer mesh, also containing magnets, to a **gravity tank** located in the upper engine room. The outlet pipe from the bottom of the gravity tank leads down to the main engine where the pipe branches to the individual bearings and gear meshes. Each bearing housing is fitted with a sight glass and thermometer for verifying adequate lubrication.

The lubricating-oil service pump is intentionally oversized beyond the bearing requirements to ensure that the gravity tank is always full. The surplus lubricating oil overflows from the gravity tank back to the sump via an overflow line, which contains a **sight glass** or **sight flow indicator** at the operating platform to visually indicate proper functioning of the system. Low-pressure cut-in switches will automatically start a second lubricating-oil service pump if the discharge pressure of the pump in service drops. In the event that the oil flow from the pump ceases, the gravity tank provides several minutes flow, so that the turbines can be stopped to avoid wiping the bearings.

The pressure lubrication system uses a pressure-regulating valve to control pressure, by recirculating surplus oil back to the sump. In this system, there is no gravity tank and the system relies completely on a standby pump to start automatically if the operating pump fails. It may be necessary for the standby pump to be battery-operated to ensure availability in the event of loss electric power.

Associated with the main lubricating-oil system are the lubricating-oil storage, settling, and purification systems (see Fig. 23.17). These systems are used for filling the system, maintaining a reserve, allowing gravity settling of heavy contaminants, and transferring oil from one system to another.

23.9 Diesel engine lubricating-oil systems

Fig. 23.18 shows representative lubricating-oil systems for a low-speed diesel engine. The lubricating-oil requirements are divided between two systems: cylinder oil, which is consumed within the cylinders and must be continually replaced; and circulating lubricating-oil, which lubricates and cools the bearings, and (usually) the pistons. The circulating oil is continuously cleaned by the lubricating-oil purifier.

The lubricating-oil circulating system draws suction from the engine sump tank, which is built into the ship's double bottom. The suction bellmouth is located slightly above the sump bottom to avoid drawing in any settled water or sediment. The oil is drawn through a suction strainer by the lubricating-oil circulating pump, and then discharged through a finer strainer and the lubricating-oil cooler. A lubricating-oil cooler bypass valve is fitted to control the temperature of the oil supplied to the engine. The lubricating oil is then fed to the bearing-oil manifolds, the turbocharger bearings, and the hydraulic governor and control circuits.

To protect the turbocharger, which takes several minutes to slow to a stop in the event of an emergency shutdown, a lubricating-oil gravity tank is installed. Although not included in the figure, storage and settling tanks are provided for the main engine oil, as in Fig. 23.17.

The cylinder-oil system is used on crosshead engines and some trunk-piston engines. The cylinder oil is high-viscosity oil used to lubricate the piston rings and cylinder liner. It has a **total base number (TBN)** high enough to provide an alkalinity that will neutralize the fuel's sulfur content. Cylinder-oil flow is controlled by engine-driven lubricators, which inject the oil in small quantities during each cycle. The oil burns off during subsequent power strokes. This system uses a measuring tank as a gravity source to the engine lubricators. The measuring tank is filled daily from cylinder-oil storage tanks.

For trunk-piston engines, cylinder lubrication is usually provided from the lubricating-oil circulating system.

23.10 Bilge systems

The bilge systems are used to remove water from the tank tops or bilges in the machinery spaces and other compartments throughout the ship. Bilge systems are divided to handle clean and oily bilge water. The clean bilge system pumps bilge water from dry cargo holds or other normally oil-free spaces including bilges in machinery spaces and the shaft alley. Clean bilge water is discharged directly overboard. However, because water in machinery space and the shaft alley bilges may be contaminated with oil, an oily-water bilge system supplements the clean bilge system in these spaces. Oily bilge water is not discharged overboard.

The clean bilge system uses manifold valves that branch to suctions in the spaces that the system serves, as illustrated in Fig. 23.19. At least two centrally located pumps, usually large centrifugal pumps, are required, but these pumps may also be used for other seawater services. On dry cargo ships the ballast system is cross-connected to the clean bilge system as shown. The manifold suction valves are of the stop-check type to prevent accidental flooding of the spaces. The clean bilge system is used as required, under manual supervision. On tankers the ballast system is divided, with only those ballast tanks aft of the engine room served by the bilge system, while the ballast system forward of the engine room is completely separated from the bilge system.

Bilge wells, also known as **rose boxes**, are recessed into the tank tops of the machinery spaces, the shaft alley, and dry cargo holds to serve as basins to collect bilge water. They are covered by perforated steel plates that serve as primary strainers. The bilge suction branch pipes draw from the bottoms of these wells. Additional strainers are fitted in accessible sections of the suction piping.

Centrifugal bilge pumps must be primed in order to operate. Central vacuum systems may be used to remove air from the pump suction piping and casing to prime the pump, or a priming unit may be fitted to each bilge pump, or self-priming pumps may be used. For clean-bilge pumps, a sea-suction valve piped to the bilge suction manifold can be cracked to flood the pump casing with seawater and thus prime the pump.

An oily-bilge system is shown in Fig. 23.20. The **oily-bilge pumps** draw from the bilge wells in the machinery spaces and the shaft alley and discharge to a **bilge holding tank**, also called the **oily-water collecting tank**. The oily-bilge system usually operates automatically, with low-capacity float-controlled pumps that start as soon as a well is almost full. In the example shown, the oily-bilge pumps are **sump pumps** comprising a vertical centrifugal pump at the bottom of the well, driven through an extended shaft by an electric motor above the bilge. Because the pump is submerged when water has accumulated in the well, no priming device is needed. An alternative is to use a single, centrally located oily-bilge pump serving all of these wells.

When the oily-water collecting tank is full, the water is pumped through an **oily-water separator**, with the cleaned water discharged overboard and the oil retained in the **waste-oil tank**. If necessary, in port or at an anchorage, the contents of the oily-water collecting tank can be transferred through deck connections to a barge.

Most troubles with bilge systems are caused by clogging with debris: clogged bilge-well covers or clogged suction strainers, or by failure of suction valves in branches not in use to completely close because of debris lodged between the disk and seat (thereby preventing vacuum formation in the suction main).

To permit pumping down high engine-room bilge levels while troubleshooting a problem system, one bilge pump is required to have an **independent bilge suction**. In addition, an **emergency bilge suction** is fitted, piped directly from an open suction pipe just above the tank top to the largest seawater pump (usually a main circulator or main-engine seawater cooling pump) as an emergency means of pumping a flooded engine room.

23.11 Ballast system

The ballast system is used to transfer seawater into and out of ballast tanks located throughout the ship. The system usually consists of centrifugal pumps that are connected to ballast system fill and suction manifolds. The pumps are not needed for all ballast operations - tanks above the waterline will drain, and tanks below the waterline will fill, by gravity. On dry-cargo ships, the ballast system is cross-connected to the clean bilge system as shown in Fig. 23.19. On tankers the ballast system is divided, with only those ballast tanks aft of the engine room served by the bilge system, while the ballast system forward of the engine room is completely separated from the bilge system.

23.12 Ship's-service compressed air and control air systems

Steamships and motorships alike are typically fitted with two compressed-air systems (see Fig. 23.21) operating at about 100 psig (7 bars), the **ship's-service compressed air system** and the **control-air system**. On diesel vessels there will also be a **starting-air system**, discussed below. Ship's-service air is piped to shops, to important locations in the engine room, and to deck, and is used for general purposes, including cleaning parts, pneumatic tools, and miscellaneous air motors.

Compressed air is stored in cylindrical steel pressure vessels called receivers, which are recharged by motor-driven air compressors. The compressors are controlled by pressure switches on the air receivers that normally cycle the compressors on and off to maintain the receiver pressure as the air is used. Generally the compressors are air-cooled, two-stage reciprocating compressors with intercooling between the stages. Often the reciprocating compressors are belt-driven by the motors, as in the figure. Sometimes rotary air compressors are used.

Intermittent operation of a compressor is appropriate because the normal demands for air are low and occasional. However, to meet high, long-running air demands it may be better for the compressor to run continuously, avoiding the frequent starting and stopping called **short cycling**, by using the pressure switches to control an **unloader**. The selection for intermittent or continuous operation is determined by the position of the **hand-off-auto (HOA)** switch where "auto" uses start-stop control and "hand" provides continuous operation with unloading. The operation of an unloader is shown in the detail of Fig. 23.21. When "hand" is selected, the solenoid admits compressed air to the unloader, forcing the fork down to hold the suction valve open: the compressor will then run without compressing air.

The control-air system supplies air for pneumatic regulating and control equipment. Clean, dry control air is very important for reliable operation of pneumatic equipment. The arrangement of the control-air system is generally similar to that of the ship's-service air system, although usually with control only by cycling of the compressor. It is common to use an **oil-free compressor**, with a **control-air dryer** at the receiver outlet to dehumidify the air. To provide a standby source of control air, the receiver may be cross-connected to the ship's-service system. Pressure-reducing and filter stations are located prior to pneumatic devices to further clean the air and adjust the pressure. These filter stations are fitted with automatic traps or petcocks for draining accumulated moisture.

23.13 Compressed air systems - motorships

Large diesel engines are started with compressed air. Regulations require two or more **starting-air receivers**, with a total stored air capacity sufficient for six starts for a non-reversing main engine, and twelve for a direct-reversing main engine. The air is stored at high pressure, typically about 450 psig (30 bars). A typical starting-air system is shown in Fig. 23.22. At least two large **starting-air compressors** are required, usually two-stage, water-cooled, motor-driven reciprocating units. A smaller **topping compressor** is often fitted to maintain a continuous charge of compressed air in the receivers. The receivers are fitted with automatic drain valves to remove condensate produced by the compression and cooling of humid air.

The starting-air receivers also supply air for starting the ship's-service diesel engines. An **emergency air compressor** fed from the emergency switchboard, with an **emergency starting-air receiver**, is fitted for cold-ship start-up of the diesel generators.

In the example shown, the starting-air receivers are cross-connected to supply the ship's-service air receiver through a pressure-reducing valve, and as a standby source of control air. Often a ship's-service air compressor is fitted for normal service and the crossover is reserved for standby use.

23.14 Firemain

The fire main is a pipe that runs through the ship, with branches to fire stations and other outlets. An example is shown in Fig. 23.23. The fire main is supplied with seawater by at least two high-capacity, medium-pressure centrifugal pumps located in the engine room. The pressure must be high enough to supply fire hoses even at the uppermost decks. The system is also used to provide **water on deck** for washing-down, to provide water for rinsing the anchor chain as the anchor is raised, and to drive bilge eductors used in lieu of bilge pumps. Typically on merchant ships, the fire pumps are kept lined up and ready to run, while naval practice is to maintain the fire main under continuous pressure. The fire pumps may be started locally, or from switches outside the engine room, or from the bridge. The rules permit some cross-connection of fire, bilge, and ballast systems. An additional fire pump may be required to be located in a watertight compartment outside the engine room.

23.15 Domestic freshwater system

Freshwater is distributed through living and working spaces for drinking, cooking, and washing. The water is supplied from storage tanks that are filled from the evaporators at sea, or from shore connections in port. Pumps draw from the storage tanks and pressurize the distribution system.

To avoid continuous pump operation, especially during periods of low consumption, the pumps discharge to a **pneumatic pressure tank**, sometimes called a **hydrophore tank**. See Fig. 23.24. This tank is partly filled with compressed air. The air pressure in the tank controls pump operation. As water is consumed, the air pushes tank water into the system, and as the water level drops, the air expands and its pressure drops. A pressure switch automatically starts the pump. With the pump running, the level in the tank rises, compressing the air until the pressure switch stops the pump.

The term **potable water** refers to water that is suitable for drinking. Often there is no need to identify the potable water separately from the wash water. However, when the freshwater is produced at temperatures too low for sterilization, as is often the case with low-pressure evaporators (see Chapter 24), the potable water is passed through a **sterilizer** prior to distribution.

Hot water is supplied from a **hot-water heater** with a steam-heating coil. To provide hot water when steam is not available an electric immersion heater is usually installed as well. Cold water is supplied to the hot water heater under pressure from the potable water system. The hot water line is a loop through which the heated water is continuously circulated to taps at washbasins and other fixtures by a hot-water circulating pump, to ensure immediate availability of hot water at each tap without waste.

23.16 Sanitary water system

The sanitary water system supplies flushing water to toilets. If seawater toilet flushing is used, the sanitary water system may have a pneumatic pressure tank like the domestic freshwater system, with the sanitary pumps taking suction from a sea chest. To avoid the high maintenance associated with a seawater flush system, many ships use freshwater flushing even though this practice increases freshwater consumption.

23.17 Tanker cargo-handling systems

Tanker cargo systems include the fill/discharge manifolds on deck, piping and valves, stripping systems, inert-gas systems, gas-freeing systems, and tank-washing systems. Complexity of the systems varies depending on the type of cargo carried, for example, a crude-oil carrier may carry only one or two grades of oil, while a product carrier may carry multiple liquids which must be segregated from each other.

Tanker cargo loading and discharge operations are under the direction of the cargo operations officer, usually the first or chief mate. The watch engineer's responsibility is the operation of support equipment in the machinery spaces. The cargo officer monitors and controls the loading and discharge from a central control room. Modern control rooms are automated with a console-mounted flow diagram integrally fitted with switches for remote-operated motorized valves, valve position-indicating lights, and remote tank-level indicators. The system is supplemented with manual soundings and monitoring of operations by crew members.

Shore-side pumps are used to load the cargo, but the ship's cargo pumps are used to discharge. On most tankers, a separate pump room is located just forward of the engine spaces. The cargo pumps, usually large centrifugal pumps, are located in the pump room, while their prime movers, usually a steam turbine for each pump, are located in the engine room. The drive shaft penetrates the separating bulkhead, which is watertight and gas tight. Pump speed may be controlled directly from the cargo-control room, or by the engineer on watch under orders from the cargo officer. Generally each pump has its own discharge line and manifold connection.

In discharging the cargo, the last of the oil in each tank is removed from the tank by **stripping** the tank. As the tank level drops, the cargo pump would eventually suck air and "lose its prime." Some tankers have positive-displacement stripping pumps for this purpose, but in the example of Fig. 23.25, **stripping eductors** are fitted. During the stripping operation, some of the cargo discharged by the cargo pumps is recirculated via the ship's slop tank to provide enough flow to drive the eductor. The slop tank is the last tank to be pumped out.

23.17.1 Cargo tank cleaning

Crude oil contains hydrocarbon sludges and waxes that tend to coat the tank bulkheads and frames as the oil is discharged. This accumulation is generally removed by **crude-oil washing (COW)** during cargo discharge. In crude-oil washing, a portion of the cargo-pump discharge flow is recirculated to **tank-cleaning machines** that are fixed in place in the tanks (see Fig. 23.25). The machines are driven by the oil pressure, and have rotating spray nozzles that operate through a cycle to spray the oil against the tank interior surfaces in a descending pattern as the tank is emptied. Because an explosive atmosphere can be created by spraying oil in the tank, the **inert-gas system** must be used during crude oil washing (see below).

On any oil tanker, including products carriers as well as crude carriers, an empty cargo tank will have to be **gas-free** if it is necessary to enter the tank for inspection or maintenance. To gas-free a tank it must first be thoroughly cleaned, and then ventilated. The thorough cleaning is done with seawater, used fixed tank-cleaning machines if installed, as well as portable tank-cleaning machines often called **Butterworth machines**. A portable machine is similar to a fixed machine in concept and construction, but is specifically designed to be lowered into the tank on the hose that supplies it, through bolted plates in the main deck. The seawater is supplied by a **tank-cleaning pump** in the engine room. The cargo stripping pumps or eductors are used to strip the accumulated water and oily waste from the tank being cleaned to the slop tank, where much of the water will separate by settling and be pumped overboard.

23.17.2 Inert-gas system

To limit the danger of explosion in cargo tanks, tankers are fitted with an **inert-gas system**, to fill the tanks with inert gas that is too low in oxygen content, at about 2 to 6%, to support combustion, which requires about 10% oxygen by volume. The source of inert gas is generally the flue gas from the ship's boilers, as in Fig. 23.26, or it may be supplied from an independent inert-gas generator. The gas is distributed to the cargo tanks to fill the void space above the cargo as the tanks are emptied and to maintain the tanks **topped up** as necessary afterwards. The tank is maintained under a slightly positive pressure of inert gas to prevent intrusion of outside air under all normal conditions: when loaded with cargo, discharging, ballasting, or loading new cargo.

Major components of the inert gas system are shown in Fig. 23.26.

The **scrubber** cools the flue gas by passing it through a shower of seawater, which also removes sulfur and particulates. At the exit of the scrubber is a demister to remove seawater droplets from the scrubbed gas.

The **inert-gas fans** draw the cleaned flue gas from the scrubber for distribution via the **inert gas main**. A pressure-regulating valve is fitted at the fan discharge to maintain the limited positive pressure in the inert-gas main. To avoid overheating of the fans when the regulating valve is almost closed, each fan discharge is fitted with a recirculation valve, which automatically opens as fan discharge pressure rises.

The **deck water seal** and non-return valve both act to prevent oil vapor from flowing back into the boiler uptakes. The deck water seal relies on trapped water and gravity to function.

The **pressure-vacuum breaker** branches off the inert gas main and serves as both a pressure relief valve and a vacuum breaker to limit structural loads in the cargo tanks. The tanks should not be subjected to more than 2-1/4 psig (1.15 atm) internal pressure or ½ psi of vacuum (0.97 atm). The pressure-vacuum breaker balances the weight of oil or antifreeze trapped in a column against the gas pressure to produce the relief mechanism.

A mechanical relief valve fitted to the inert gas main provides further protection against high internal pressure. This valve discharges into the **mast vent** or **mast riser**, located forward, clear of the principal living and working spaces. The outlet has a high-velocity discharge cone that vents toxic inert gas at a high velocity to reduce personnel exposure. The outlet is at least two meters above the deck level, and contains a flame-arresting screen to prevent a deck fire from entering the cargo tanks.

The inert gas system can be used to help gas-free the tanks, using a normally blanked fresh-air intake at the fans.

CHAPTER 24
ELECTRIC SYSTEMS

24.1 Fundamentals of electricity

Electricity is associated with the accumulation or the flow of **electrons**, which are negatively-charged, subatomic particles. Static electricity is caused by the accumulation of electrons, as might occur from rubbing hard rubber on a wool carpet. The resulting **static charge** of the stored electrons might then be released suddenly as a spark. In a more useful form of electricity, the electrons flow through wires to energize light bulbs and computers and to turn the shafts of motors.

As long as a complete electric path or **circuit** is available, electron flow can be initiated and sustained in a number of ways, including chemical reaction and magnetic induction. **Batteries** and **fuel cells** cause electron flow by chemical reaction. For most applications **alternators** or **generators** are used to produce electron flow by magnetic induction, in which the mechanical power output of an engine is used to drive conductors through a magnetic field, **inducing** an electron flow in the conductors. Electron flows can also be created by **thermocouples** and **photovoltaic cells**, which respond to heat and light, respectively. Thermocouples are commonly used to generate electric signals for temperature measurement. Fuel cells and photovoltaic (or solar) cells have been used in low-power applications for years but are becoming more common and more capable of higher-power applications.

Electrons flow from areas of higher electron concentration (negative charge) to areas of lower electron concentration (less negative charge) if a conductor path is established. The conductor can be solid, like a copper wire, but liquids, such as salt water and mercury, and ionized gases, can also conduct electricity. The effects of electron flow are recognizable and include, as examples, heating of the conducting wire element of an electric heater, causing a magnetic field around the conductors in an electric motor to produce mechanical torque, and heating of the ionized gas in a fluorescent lamp to produce light.

24.2 Electrical terms

Engineers do not normally speak of flowing electrons but rather of electric **current**. Current is an imaginary stream that flows from the less negatively charged area (more positive) to the more negatively-charged area (less positive). That is, the direction of the current is opposite to the electron flow. This convention was adopted because it is easier to think of a flow going from a higher level (more positive) to a lower level (less positive). Current flow is normally given the symbol **I** and is measured in units of **amperes**, usually abbreviated to **amps**. Current in amperes is analogous to fluid flow in gallons or liters per minute.

The force used to push current through a conductor is the **electromotive force (emf) or potential**. It is measured in units of **volts** and identified by the symbol **V** or **E**. The fluid analog of the electromotive force in volts is pressure in psi or pascals.

24.2.1 Direct current vs. alternating current

Electricity may be supplied as either **direct current (DC)** or **alternating current (AC)**. DC flows in only one direction in the conductor because the potential always acts in the same direction. Current originating from a battery in a flashlight and flowing to the bulb is an example of DC. The current flows from the positive terminal of the battery, through a circuit consisting of the switch, the bulb, and the conductors, to the negative terminal of the battery. Only the resistance (to electron flow) of the circuit limits the current.

With alternating current the driving potential (or voltage) continually reverses direction, forcing the current in the circuit to continually reverse direction as well. The voltage takes the form of a sine wave as shown in Fig. 24.1, rising first to a positive peak value as shown, then falling through zero to an equal negative peak value. As a result, the current first flows in one direction, peaks, slows, stops, reverses direction, increases to an equal peak value in the other direction, slows, stops, and resumes its flow in the original direction. In the United States the standard generating systems produce alternating current that reverses with a **frequency** of 60 **cycles per second (cps)**, commonly referred to as 60 **cycles** or 60 **Hertz** (**Hz**). In Europe and Japan, the standard is 50 Hz.

24.2.2 Resistance and resistance "loss"

Resistance to current flow is symbolized by **R** and is measured in units of **ohms**. Resistance can be calculated by **Ohm's Law**: if E is the voltage and I is the current,

$$R = E/I$$

The magnitude of the resistance of a conductor is affected by the ease with which electrons (current) flow in the material. If electrons flow easily the material is said to have low resistance. Copper is a better conductor than steel, and is used in most power applications. The resistance is also depends on the available area through which the electrons flow, and on the conductor length. As with fluid in a pipe, the narrower or longer the path, the more difficult it is to push the electrons through. Since most electric conductors are round, so that area is proportional to diameter squared, conductor area is often abbreviated to **circular mils**, where a mil is 0.001 inch, and a circular mil is the diameter squared. As an example a wire of 0.100 inch (100 mil) outside diameter is said to be a 10,000 circular-mil wire. Engineers often identify a conductor by its **American Wire Gage** (**AWG**) number. In this system the higher the number, the smaller the diameter, so that #12 wire is thinner than #10 wire.

The unit **megohm** is a resistance of one million ohms and is used to express the resistance of electric insulation.

Just as friction and other resistance to fluid flow in piping results in a "loss" of energy, so does the resistance in a conductor result in a "loss". This "loss" is proportional to the square of the current, multiplied by the resistance:

$$\text{resistance "loss"} = I^2R$$

Because it is a function of the current squared, a modest increase in current in a conductor results in a much larger "loss" of energy. This "loss" manifests itself as heat.

The unit **megohm** is a resistance of one million ohms and is used to express the resistance of electric insulation.

24.2.3 Magnetic field

An electron flow, or current, is always surrounded by a magnetic field. When a conductor is wound into a cylindrical coil, the magnetic field that will be formed will be oriented along the axis of the cylinder. One end of the coil will act like the north pole of a permanent magnet, and the other end, like the south pole. The coil is therefore an **electromagnet**.

24.2.4 Impedance

With alternating current, the electron flow is opposed not only by the resistance but also by an additional factor, the **reactance**. Reactance is caused by the response of the circuit to the cyclic fluctuations of the alternating current and voltage, and may be further categorized as inductive reactance or capacitive reactance. The sum of the resistance and the reactance is the **impedance** (**Z**), and has units of ohms. Impedance is defined as:

$$Z = E/I$$

The magnitude of the impedance depends on factors that include the characteristics of the components in the circuit and the frequency of the current.

24.2.5 Open circuits, short circuits, and grounds

An **open** in an electric circuit occurs where the conducting path has been broken, stopping the flow of current. An open occurs when a switch is opened, or a wire breaks or becomes disconnected.

A **short** circuit occurs when the current path is bridged, allowing the current to bypass one or more components, for example, when two conductors accidentally touch.

A **ground** occurs when an electric circuit is accidentally or intentionally connected to a very low potential sink, like a ship's hull or the sea. A common example of a ground occurs when insulation on a conductor breaks down, allowing the conductor to contact the adjacent metal housing or structure.

24.3 Electric power

The energy of flowing electrons is a power source that can be used do work in an electric motor or can generate heat as in an a toaster element. The unit of power associated with electric devices is the **watt** (**W**), which is the power produced when one ampere of current is pushed through an electric circuit by one volt of electromotive force (a potential difference). In a similar manner, hydraulic power is developed when a flow of fluid is pushed through a fluid circuit by pressure difference. Since the watt is a relatively small amount of power, **kilowatts** (**kW**) and **megawatts** (**MW**) are commonly used. One kW is equal to 1000 W, and one MW equals 1000 kW.

The law of conservation of energy dictates that electric energy or power can be converted to mechanical power or to heat, and vice versa. If the conversions were complete, one kW of electric power would provide a kW of mechanical power or a kW of heat. In familiar units, these approximate energy equivalents are listed below:

$$746 \text{ W} = 0.746 \text{ kW} = 1 \text{ hp}$$
$$1 \text{ kW} = 1.340 \text{ hp}$$
$$1 \text{ kW} = 3412 \text{ Btu/hr}$$

24.3.1 Power from direct current

In a DC circuit, where the voltage and current are at constant, steady values, the power in kW can easily be calculated as the product of voltage and current in amps:

$$kW = E \times I$$

24.3.2 Power from alternating current

In an AC circuit the power cannot be calculated so easily because first, both voltage and current fluctuate in magnitude and direction in sinusoidal patterns, and second, because the voltage and current do not necessarily act in unison, as shown in Fig. 24.1. To account for the cyclic variation in magnitude of voltage and current, an averaged value is used, called the **root-mean-square** or **rms** value. See Fig. 24.3. For AC circuits the values typically cited for current (I) and voltage (E) are in fact the rms values, and not the peak values. These rms values are each about 71% of the peak values. For 120 volt, 60Hz power, the rms values are +120 volts and -120 volts, while the peak values are +170 volts and -170 volts, yielding a **peak-to-peak voltage** of 340 volts.

In Fig. 24.2, note that the voltage reaches its peak value before the current reaches its peak, so that the current is said to be **lagging**. If the current were to reach its peak first we would say that it is **leading**. The degree of lagging or leading is dependent on the circuit components, and their response to the cyclic fluctuations of the alternating current and voltage. Note that it takes 360 degrees for a sine wave to complete its cycle, enabling the abscissa in Fig. 24.2 to be given in degrees as shown. The difference between corresponding points on the two sine waves for voltage and current, measured in degrees, is called the **phase shift.**

Power is the product of current and voltage, but it can be seen in Fig. 24.2 that whenever there is a phase shift, this product will never reach the high values that it would reach if the current and voltage acted in unison. For example, at 90 degrees the voltage is 100% of its peak value but the current is at only 60% of its peak value, so that the instantaneous power is only 60% of what it would be if current and voltage reached their peak values together at 90 degrees. To account for the lower power caused by a phase shift, the **power factor** (**pf**) is defined as follows:

$$\text{pf} = \frac{\text{active power}}{\text{EI}} = \frac{\text{active power}}{\text{apparent power}}$$

The **apparent power** is simply the product of the rms voltage (E) and the rms current (I), and is expressed in volt-amperes (VA) or kilovolt-amperes (kVA). The **active power**, which is the **real power**, is expressed in watts or kilowatts.

$$\text{active power} = \text{real power} = \text{pf} \times \text{EI} = \text{kW}$$
$$\text{apparent power} = \text{EI} = \text{kVA}$$
$$\text{kW} = \text{pf} \times \text{kVA}$$

If there is no phase shift, then the voltage and current follow sine curves that are exactly in step. In these cases the voltage and current are said to be **in phase,** the active power is equal to the apparent power, and the power factor is one. This condition is the exception, rather than the rule. More often, the pf is less than one because the voltage and current cycles do not coincide, that is they are **out of phase**, and the active power (kW) is therefore less than the product of volts and amperes. The significance of the power factor, therefore, is that only at a power factor of one, where the voltage and current are in phase, will the real power be equal to the product of the volts and amperes. Usually the voltage and current are out of phase, the pf is less than one, and the active power is less than the product of volts and amperes.

To appreciate the significance of having voltage and current out of phase, it should be noted that what is required from an electric circuit is usually a certain amount of real power, such as might be needed to drive a motor. If the power factor is less than one, and if the voltage is fixed, then for the required real power a higher current is necessary than if the voltage and current were in phase. Recall that whenever there is a current in a conductor there is also a "loss", and that this "loss" is proportional to the square of the current. For example, in a circuit supplying a motor with AC power, if the pf is 0.8, then the current must rise to 125% of what would be required for the same power if the pf were one, while the "loss" in the conductors rises to 156% (125% squared). Viewed another way, all of the current contributes to wasteful heat generation, even when only a portion of the current contributes to the real power.

24.4 Basic electric circuits

Basic electric circuits are either **series** circuits or **parallel** circuits.

In a series circuit, the components are connected in succession, so that the total line current must pass through each of the components in turn, as in Fig. 24.4. Therefore, for the resistors in this circuit:

$$I = I_1 = I_2 = I_3$$

The total voltage drop across these resistors is the sum of the voltage drops across each of the resistors:

$$V = V_1 + V_2 + V_3$$

Recalling Ohm's Law (I=V/R or V=IR), the **equivalent resistance (R_e)** of the series circuit is:

$$R_e = R_1 + R_2 + R_3$$

And the voltage drop across each of the resistors is equal to the line current times that resistance of that resistor:

$$V_1 = I_1R_1 \qquad V_2 = I_2R_2 \qquad V_3 = I_3R_3$$

In a parallel circuit, the components are connected in branches, which permit the current to divide and flow through two or more independent parallel paths, as shown in Fig. 24.5 on the left. Therefore, each of the three resistors has the full line voltage across it:

$$V = V_1 = V_2 = V_3$$

At the same time, the total line current is the sum of the three branch current flows:

$$I = I_1 + I_2 + I_3$$

The equivalent resistance of a parallel circuit is the value of a single resistor that would pass the same line current as the group of resistors in the parallel circuit, as shown on the right in Fig. 24.5. This equivalent resistance can be calculated for the parallel circuit of the figure using Ohm's Law, which yields:

$$R_e = \frac{1}{1/R_1 + 1/R_2 + 1/R_3}$$

24.5 Electric components and instruments

A short list of electric components and instruments most likely to be of interest to marine engineers appears in this section.

Voltmeters are used to measure the strength of the electromotive force or potential between two points.

At the heart of a voltmeter is a pivoted movement, surrounded by a coil. The movement is attached to the pointer, and turns in response to the magnetic field produced by the small current that flows in the coil when a voltage is applied across the meter terminals. Voltmeters are connected in parallel to the component for which the voltage drop is desired as shown in Fig. 24.6. Voltmeters must have high internal resistance so that the current they draw is negligible. In low-voltage electronic circuits, ordinary voltmeters may draw too much current to permit their use.

Voltage testers are used where it is necessary to know only whether there is full voltage or no voltage.

Ammeters are used to measure current. The basic movement of an ammeter is similar to that of a voltmeter, but the coil of the meter is wired in parallel with a low-resistance resistor called a **shunt**. The movement then responds to the voltage drop associated with a particular current in the shunt. The scale is graduated in appropriate units such as milliamperes (1 ma is one-thousandth of an amp) or amps, depending on the shunt. Ammeters are connected in a series with the conductor in which the current flow is to be measured. A basic ammeter with its integral shunt is shown in Fig. 24.7.

Clamp-on ammeters are hand-held instruments used with AC circuits. A plier-like pick-up coil is closed around the conductor in which current is to be measured, and the movement responds to current induced in the pick-up coil by the changing magnetic field in the conductor.

Ohmeters are used to measure resistance. An ohmeter consists of a simple meter movement connected in series with a battery. When the ohmeter is connected in series with the component or circuit to be tested, the battery, which acts as a constant-voltage source, supplies current to develop a voltage drop across the test device. The movement is calibrated to read in ohms. **Ohmeters must never be used on energized circuits.**

A **megger** is a special ohmeter used for testing electric insulation. The voltage source is a hand-cranked DC generator. Meggers are used only when equipment is secured and isolated from the power source. After connection to the selected insulated component, the high voltage from the generator is applied momentarily, while the meter measures the current flow through the insulation. Meggers are calibrated in megohms.

A **multimeter** is an instrument containing one movement that can be connected by a selector switch to a variety of shunts to serve as an ammeter, or to special resistors to serve as a voltmeter, or to a small dry-cell battery for use as an ohmeter.

Wattmeters are used to measure electric power. Since electric power is the product of current and voltage, a wattmeter will have four connections, two for measuring current and two for measuring voltage.

Clamp-on wattmeters are similar to clamp-on ammeters in having a plier-like pick-up coil to measure current, but they also have two or three leads for connection to the power leads, in order to simultaneously measure voltage.

Watt-hour meters are used to measure the amount of energy consumed over time. Watt-hour meters must not only measure power (current and voltage) but must also have device for measuring the time interval involved.

24.6 Alternators and generators

Generators are machines for producing DC electric power, while alternators are machines that produce AC power. It is common to refer to alternators as generators. Aboard ship generators and alternators are rotating machines driven by engines or turbines, serving as the normal source of electricity, and on larger vessels, for standby and emergency electric power as well.

When a conductor is passed through a magnetic field, or when a magnetic field is passed over a conductor, an **electromotive force** or **impressed voltage** is induced in the conductor. If the conductor is part of a complete circuit, the current that will then flow is called an **induced current**. The linkage of relative motion between a magnetic field and a conductor, and the resulting impressed voltage and induced current, is the principle behind generators and alternators. In a generator, conductors are arranged on the rotor, driven through a stationary magnetic field by a prime mover, with the induced current collected from the rotor for delivery to the load. In most alternators, the conductors are stationary, and surround a rotating magnetic field: current is induced in the stationary conductors as they are swept by the magnetic field.

24.6.1 Single-phase alternators

In the simple alternator of Fig. 24.8 the engine or turbine spins a magnet (the **rotor**), with north and south **poles**, which is surrounded by a coil of stationary conductors (the **winding** on the **stator**). The magnitude of the impressed voltage will depend on the strength of the magnetic field, denoted as **magnetic flux**, and the speed at which the conductors are swept or "cut" by the magnetic flux.

Because of the rotation of the magnetic poles the impressed voltage is not constant, but varies in magnitude and direction cyclically, as shown in Fig. 24.1. The flux cutting the right side of the conductor is that associated with a south pole, but when the magnet rotates 180 degrees the conductor is being cut by the flux associated with the north pole, reversing the direction or **polarity** of the voltage. When the magnet is vertical no flux cuts the conductors and therefore no voltage is impressed. The result is a voltage curve with the shape of a sine wave, as shown in Fig. 24.1. Since this alternator generates voltage that is described by a single sine wave it is a **single-phase** alternator.

If the two-pole rotor (one north and one south pole) of Fig. 24.8 is rotated at 3600 rpm the current will have a frequency of 3600 cycles per minute or 60 cycles per second. If the rotor were a four-pole magnet as in Fig. 24.9, then in each revolution of the rotor the conductors in the winding will be cut by the magnetic field twice as frequently. Turning the rotor at 1800 rpm would then produce the same 60 Hz sine wave. The frequency of the voltage generated by an alternator can be expressed as:

$$f = \frac{rpm \times N}{2 \times 60}$$

where f is the frequency in Hz and N is the total number of poles.

24.6.2 Three-phase alternators

If a two-pole generator were fitted with three sets of windings on the stator, as in Fig. 24.10, a three-phase alternator results. The windings are positioned 120 degrees apart and are not connected to each other. As the poles of the field rotate a sinusoidal voltage is induced in turn in each winding, 120 degrees out of phase with the voltage produced in the adjacent windings. Each winding has one end connected to a common ground, and the voltages that result, V_1, V_2, and V_3, measured relative to ground, are shown in Fig. 24.11.

24.6.3 Field and excitation

In most generators the magnet that provides the field is an electromagnet, comprised of a steel **core** wound with coils of wire that form the **field windings**, as in Fig. 24.12. The **field** is energized by the

excitation current that flows through these field windings. To maintain the polarity of the field, the excitation must be direct current. Fig. 24.12 shows how copper **slip rings** mounted on the rotor are used with stationary carbon **brushes** to provide the electrical connection from the excitation source to the field windings. The brushes are pressed against the slip rings by springs, and are replaced as they wear.

Most generators are **self-excited**, meaning that the excitation current is derived from the power produced by the generator. Starting of a self-excited generator makes use of the residual permanent magnetism of the field core. Once the generator starts rotating, the residual permanent magnetism induces current to begin to excite the field coils, and the resulting magnetic flux builds up to the desired level.

24.6.4 Direct-current generators

Since any generator with a rotating element will naturally produce an alternating voltage, it is necessary, in a DC generator, to continually convert this alternating voltage into a direct voltage. In traditional DC generators the construction of the machine is reversed relative to the alternators discussed above, so that the magnetic field is stationary, while the windings rotate, as in Fig. 24.13. The rotor is then called the **armature**, and is fitted with a rotating switching device called a **commutator**, which does the AC to DC conversion. The voltage produced in the windings of the armature is sinusoidal, as it is in an alternator. If this voltage were taken out of the rotor by slip rings and brushes, as in Fig. 24.12, single-phase AC would be delivered to the generator terminals. The commutator is essentially a single slip ring divided into separate segments called **commutator bars**. Each terminal of the armature winding is connected to one commutator bar. As the armature rotates, the winding terminal beneath each brush switches half way through every revolution, simultaneously with the change in polarity of the voltage in the winding. The polarity of one brush and its connected generator terminal is thus always positive, while the other brush and generator terminal is always negative.

The DC voltage produced by a generator with a single armature winding and a two-bar commutator fluctuates as shown in the upper plot of Fig. 24.14. To obtain a more constant voltage more windings would be added to the armature, with each winding connected to a pair of diametrically-opposed commutator bars, to yield an output voltage curve as shown in the lower plot of Fig. 24.14. The more windings there are on the armature, the steadier will be the resulting terminal voltage of the generator. Large DC generators have dozens of armature windings and commutator bars.

A commutator is only one way to achieve the continual switching necessary to convert AC to DC: electronic devices called **rectifiers** are also used. Modern rectifiers are built around solid-state **diodes**. A diode has low resistance to current flow in one direction and very high resistance to current flow in the other direction. Fig. 24.15 depicts a rectifier circuit. By appropriately connecting rectifiers to the stator windings of an alternator, DC can be provided without the maintenance problems associated with brushes and commutators. Alternators can be used in automobiles, where the power is distributed as DC, because they are fitted with rectifiers.

24.7 Electric motors

An electric motor has a role opposite to that of a generator, in that a motor draws electric power and converts it to mechanical power. Most shipboard motors are AC, because they are smaller and less costly than DC motors of similar power output. DC motors are used only rarely on modern ships. **Universal motors** operate on either AC or DC.

24.7.1 Alternating-current motors

In concept and construction, a three-phase AC motor is very similar to a three-phase alternator, as shown in Fig. 24.10. When three-phase AC current (the **line** current) is applied to the stator windings of the motor, with one phase connected to each winding, the current will produce a magnetic field around each winding that varies in magnitude and direction in a sinusoidal pattern. The magnetic field produced by each winding interacts with that of the adjoining windings to produce a net magnetic field that revolves around the stator. If a compass needle were placed in the center of the stator it would rotate at a speed coinciding with the frequency of the applied current, since the compass needle is a two-pole magnet. The direction of rotation of the field will depend on the order in which the windings are connected to receive the line current: if two of the three line connections are interchanged at the motor terminals, the direction in which the magnetic field produced by the stator windings will reverse, as will the compass needle.

The stator windings of the three-phase machine in Fig. 24.10 are spaced at 120-degree intervals, so that the field produced by these stator windings rotates once per cycle of the applied current, a speed called the **synchronous speed**. This stator is said to have two poles, and the corresponding synchronous speed can be calculated from the following relation:

$$\text{rpm} = \frac{120 \times f}{N}$$

where N is the number of poles on the stator and f is the line frequency. A two-pole stator supplied with 60 Hz line current would have a synchronous speed of 3600 rpm. In this example, the magnetic field produced by the stator and the compass needle would both rotate at this speed.

Additional windings can be fitted to the stator for each phase. Each additional winding per phase provides an additional pair of poles. For example, the stator could be arranged with nine windings, with every third winding connected to the same phase, so that there are three windings per phase. Since each set of windings provides one pair of poles this would be a 6-pole motor. The synchronous speed of a 6-pole motor, when supplied with 60Hz line current, is 1200 rpm.

24.7.2 Synchronous motors

If we replace the compass needle in the above example with a more powerful magnet, the magnetic field produced by this magnet would **lock** onto the rotating magnetic field produced by the stator windings. The rotor can be a strong permanent magnet or an electromagnet supplied (excited) by a DC source. This type of AC motor is called a **synchronous motor** and is characterized by the fact that the rotor rotates at the same speed as the magnetic field produced by the stator windings, the synchronous speed. Synchronous motors range in size from small motors for electric clocks fitted with permanent-magnet rotors, to very large propulsion motors, developing 70,000 HP or more, with electromagnetic rotors.

24.7.3 Induction motors

A more common type of AC motor, widely used for driving pumps and fans, is the **induction motor**. In most induction motors the rotor is built up of laminated steel plates, into which are inserted a number of **conductor bars**, connected at the ends by **end-connecting rings** as shown in Fig. 24.16. Without the core, the conductor bars and end-connecting rings would form a rotating cage-like structure, for which reason this type of motor is often called a **squirrel-cage motor**.

If a squirrel-cage rotor is placed in the rotating magnetic field of an AC-motor stator, but is locked to prevent its rotation, then as the rotating magnetic field, produced by the stator, crosses the conductor bars it will induce a voltage in these bars, and therefore a current will circulate through the rotor, since the bars are connected together to form a complete circuit. The current in the bars of the rotor produces a magnetic field surrounding the bars, which causes the rotor to act as if it were a permanent magnet or electromagnet. If the rotor were then released it would rotate as any magnet would, and will accelerate towards the synchronous speed of the stator field. However, it will never reach synchronous speed because, if it did, the conductor bars of the rotor would no longer be crossed by the rotating magnetic field of the stator and no induced rotor current or rotor magnetism would result. Therefore, induction-motor rotors always run at a speed slightly less than the synchronous speed in order to induce a voltage and current in the rotor conductor bars. The difference from synchronous speed is the **slip** and is defined as follows:

$$\%\ \text{slip} = \frac{\text{synchronous speed} - \text{actual rotor speed}}{\text{synchronous speed}} \times 100$$

As the torque on an induction motor increases, the rotor must slow down (slip must increase) so that sufficient current is induced in the conductor bars of the rotor to develop an induced magnetism strong enough to develop the necessary torque. However, it is important to note that the slip is not much more than five percent even at high torque, so that induction motors are almost constant-speed machines when supplied with constant-frequency line power.

As with synchronous motors, induction-motor stators can have multiple pairs of stator poles for different speed requirements, and that they can easily be reversed by interchanging two of the three line connections.

24.7.4 Single-phase motors

Although most large motors aboard ship are supplied with three-phase power, smaller motors, such as those driving appliances, must operate from single-phase circuits. **Single-phase motors** are most often arranged with two sets of stator windings, with the second set supplied from the line connections through a **capacitor**. The capacitor imposes a phase shift on the line current reaching these windings, to make it look like two-phase AC current. The stator is thus able to develop a rotating magnetic field to drive a rotor.

24.7.5 DC motors

In construction, DC motors are similar to DC generators. The traditional advantage that DC motors have had over AC motors is that, unlike AC motors, their speed can be easily controlled over a wide range by a **rheostat**. A rheostat is an easily adjusted variable resistor that can be connected in series with the motor field windings to control field current and, therefore, motor speed. However, solid-state electronic controls like variable frequency converters are now available, allowing AC motors to be operated with variable speed. Consequently, few DC motors are found aboard modern ships.

24.8 Transformers

An advantage of AC power is that **transformers** can be used to easily and efficiently change the magnitude of the voltage to best meet requirements. Transformers are simple, have no moving components, and are easily maintained.

Transformers convert AC power from one voltage to another by taking advantage of the varying magnetic field developed by an AC current and the fact that this fluctuating magnetic field can induce a voltage in a second conductor. Simple transformers consist of a **primary winding** wound around a laminated-steel **core**, which is also wound with a second winding, the **secondary winding**. Typically the core and windings are arranged as shown in Fig. 24.17. If the secondary winding has more coils or **turns** than the primary winding, the secondary voltage will be higher than the primary voltage and the transformer is referred to as a **step-up** transformer. Conversely, if the secondary winding has fewer turns than the primary winding the secondary voltage will be less than the primary voltage, and the transformer is a **step-down** transformer. The relationships between transformer voltages and currents are as follows, where n is the number of turns in each winding:

$$V_1n_2 = V_2n_1$$
$$I_1n_1 = I_2n_2$$

Transformer cores are made of steel laminates which are electrically insulated from each other in order to reduce **hysteresis** losses in the magnetic core. Overall transformer efficiencies are very high, 98 to 99%.

Autotransformers have a core with only one winding, arranged with one or more **taps**, as shown in Fig. 24.18. The taps are used to take off secondary voltages of lower magnitude than the full secondary voltage. The magnitude of the tapped voltage is dependent on the relative number of turns in the respective sections of the transformer.

24.9 Motor starters

24.9.1 Motor starting current

When the rotor of an AC motor rotates, the magnetic field surrounding the rotor interacts with the stationary windings to induce a voltage in the windings. This induced voltage is opposite in polarity to the line voltage, and therefore it actually counteracts the effect of the line voltage at the motor terminals. An analogous effect occurs in DC motor armatures, where the induced voltage is called the **counter** or **back electromotive force**, or **back emf**. The induced voltage or back emf limits the current flow through the windings.

Since a motor rotor needs to be rotating to generate induced voltage or back emf, a motor which is just being started, whose rotor has not yet begun to turn, will have no such induced voltage or back emf. At starting, therefore, current will be limited only by the impedance and resistance of the motor windings. As a result, when full line voltage is applied to a motor at starting the **motor starting current** will be many times greater than the full-power running current. As the rotor accelerates, the induced voltage or back emf increases, and the current in the line and motor windings falls to normal values.

High starting currents have two negative effects, which affect large motors in particular. First, the high current in the windings can generate enough heat to damage the motor, especially if the driven load accelerates slowly. Second, the very high starting current can draw down line voltage sufficiently to cause lights to dim and other equipment being supplied from the same line to malfunction. For these reasons all DC motors and large AC motors are started with reduced voltage applied to the motor terminals by a **motor starter.**

24.9.2 Motor starters

AC motor starters are of two general types, **across-the-line starters** and **reduced-voltage starters**.

Across-the-line starters are commonly used for small and medium-power AC motors where the line capacity is sufficient to prevent objectionable voltage drops upon starting, and where the driven loads allow the motor to accelerate quickly. As the name implies, this type of starter simply closes the necessary **contacts** to apply full line voltage to the motor terminals. Fig. 24.19 depicts a simple across-the-line starter. When the run button is depressed the magnetic-hold solenoid is energized to close the L_1, L_2, and L_3 contacts, sending current to the motor.

Reduced-voltage starters are used for larger motors or where the supply lines are not large enough to handle the full starting current. Reduced-voltage AC starters use autotransformers with one or more taps to obtain the necessary reduced voltage. A simple reduced-voltage starter is shown in Fig. 24.20. Depressing the start button initially supplies current to the motor at reduced voltage, but also starts a timer. As the motor accelerates, the timer bypasses the autotransformer to apply full line voltage to the motor terminals.

The simple starters shown in these figures include overload trips, but do not include other protective devices that are normal features of actual starters.

DC motor starters use resistors to limit the initial starting current. As the motor accelerates the resistors are automatically removed from the circuit.

24.10 Batteries and fuel cells

Batteries and fuel cells are electro-chemical, direct-current power supplies. Batteries provide power for finite periods of time before they are replaced or recharged. Aboard ship, batteries are used as power supplies for portable equipment, as temporary emergency-power sources, and for starting small auxiliary engines. On the other hand, fuel cells are continuous power sources, and while not yet in use aboard ship, they are considered a likely, near-future alternative to generators.

24.10.1 Batteries

The chemical energy stored in a battery is finite in quantity, and is measured as its **ampere-hour capacity**. As electric power is drawn from the battery, the chemical energy is depleted until the cells are discharged. Batteries are classified as either **primary** or **secondary** depending on whether they are discarded after discharge or are rechargeable, respectively. Most batteries used aboard ship are rechargeable after use.

Dry cells contain an **electrolyte** in a solid or semi-solid state, while **wet cells** contain a liquid electrolyte. Dry cells are commonly used for low-power purposes, as in flashlights and other portable equipment, and as back-up supplies for computers. Large, wet-cell **storage batteries** are used where higher power is required, such as for starting small auxiliary engines, and as short-term power supplies for emergency lighting and control circuits.

The two most common types of wet-cell storage batteries are the **lead-acid** and the **nickel-cadmium-alkali** (**Nicad**) types. Generally, a battery is made up of a number of cells connected in series, with each cell delivering 1.5 volts or less. The state of charge of a battery is determined differently for each type. For lead-acid batteries the charge is determined by measuring the density of the electrolyte with a hydrometer, while the charge of the Nicad batteries is determined by measuring the cell voltage.

24.10.2 Fuel cells

Fuel cells are similar to batteries but supply power continuously because the electrolyte is continuously rejuvenated. This rejuvenation is achieved by continuously adding hydrogen and oxygen to the electrolyte. Fuel cells operating on pure hydrogen have been used in space vehicles for years, but advancing technology will allow fuel-cell systems to operate on less expensive, readily available hydrocarbon fuels from which the hydrogen can be separated. Fuel-cell systems have the potential to achieve efficiencies higher than those based on engine-driven generators.

24.11 Shipboard electric systems

Most American ships use 60 Hz AC supplied at 480 volts. For large shipboard electric plants, where the ship's generating capacity is above 2500 kW, it is often more economical to supply the power at 2400 or 4160 volts. 50 Hz is common on foreign ships. Some naval vessels use 400 Hz for special applications.

Regulations normally require that a ship's electric-generating plant must be capable of supplying the design electric load with one generator shut down. Most ships therefore have two or three **ship's-service generators**. The ship's-service generators are connected to the **main switchboard**, which includes generator panels and **distribution switchboards**, connected by large conductors called **bus bars**. From the distribution switchboards the power is passed through **feeders** to other switchboards called **load centers, group-control centers, and power panels**. These switchboards house **circuit breakers** and control equipment known collectively as **switchgear**. Fig. 24.21 is a **one-line diagram**, which shows the **electric-distribution system** for a typical merchant ship. An **emergency switchboard** is used to supply electric power to vital loads and is normally supplied with power from the main switchboard. In the event of a complete loss of electric power (a **black-out**) the emergency switchboard is isolated from the main switchboard, and supplied with power from the **emergency generator**. The emergency generator is required to be located outside of the engine room and above the ship's margin line (3 inches or 75 mm above the bulkhead deck, sometimes referred to the upper deck).

Most shipboard motors driving pumps, compressors, and blowers are 460 volt, three-phase motors. The motors are assumed to receive power at 460 volts, although the power is generated at 480 volts, because of the **voltage drop** associated with the line losses in the conductors and switchgear. When power levels are high, high voltages (2400 or 4160 volts) are preferable for transmission since, for a given power, the current decreases proportionally as the voltage goes up (power is proportional to the product of voltage and current). Low currents allow smaller conductors and cause lower line losses (line losses are proportional to the square of the current).

For safety reasons and for commonality with shore-side equipment, the power used for lighting circuits, and for general distribution to accommodation and work areas of a ship, is reduced to a lower voltage by transformers. On American ships 120 volt, single-phase power is normal for these purposes, but 220 volt, single-phase power is common on other vessels. Specialized equipment aboard ship, such as control systems and battery chargers, usually use even lower voltages, with 24 volts being a typical value.

24.12 Electric propulsion

Electric drives for ship propulsion have been used for over 100 years, employing generators driven by steam turbines, diesel engines, or gas turbines to power the **propulsion motor**(s) on the propeller shaft(s). In comparison to geared drive, electric drive is more expensive to install, heavier, and less efficient. Electric drive is therefore normally used only where these disadvantages are outweighed by the advantages, which include excellent control of propeller speed and direction, as well as the flexibility of locating the generators and their prime movers in different spaces than the propulsion motors. A further advantage is that the power output from any number of generators can be combined to drive one or more propulsion motors.

When ship's service power as well as propulsion power is derived from the same generators, the ship is said to have an **integrated** power plant.

Fig. 24.22 shows a modern electric-propulsion system using AC/AC variable-frequency drive. In this system the mechanical prime mover drives an AC generator at constant speed, enabling ship's services to be supplied at constant frequency. For full speed of the ship, the maximum rpm of the propeller can be achieved without the frequency converter, based on the ratio of numbers of poles:

$$rpm_{motor} = rpm_{alternator} (N_a/N_m)$$

where N_a and N_m are the numbers of poles in the alternator and propulsion motor respectively. As an example, if the prime mover is a gas turbine running at 3600 rpm, driving a two-pole alternator, and the propulsion motor is a 60-pole synchronous motor directly connected to the propeller shaft, the propeller will turn at 120 rpm. (If an induction motor is used, the propeller rpm will be slightly lower.) For maneuvering, lower propeller rpm is achieved without changing the generator speed by using the frequency converter to reduce the frequency of the power to the propulsion motor. Reversing of the propulsion motor is easily accomplished by switching two of the three leads supplying the motor.

CHAPTER 25
REFRIGERATION SYSTEMS

Refrigeration is the process in which a space is cooled below the temperature of its surroundings. Considering that heat normally flows from hot to cold, it may appear that this process violates physical laws, but common refrigeration cycles are able to achieve the task by doing work on a fluid (the **refrigerant**) as it is forced through a series of thermodynamic processes and circulated through the cold space.

25.1 Refrigeration cycle

Most refrigeration systems use the vapor-compression cycle, shown schematically in Fig. 25.1. The four essential components are the **compressor**, the **condenser**, the **expansion device**, and the **evaporator**.

With reference to Fig. 25.1, the refrigerant is a pressurized liquid at point 4. When the refrigerant flows through the expansion device its pressure drops, and some of it evaporates, reducing its temperature at point 1 to a value below that of the refrigerated space.

The evaporator is installed in the refrigerated space. From point 1 the cold, low-pressure liquid refrigerant flows through the evaporator, where the remaining liquid refrigerant evaporates, absorbing the required **latent heat of vaporization** from the space. The refrigerant leaves the evaporator as a vapor at point 2.

The compressor draws the cold, low-pressure vaporized refrigerant from the evaporator at point 2, and raises its pressure, discharging it at point 3 as a high-pressure vapor. The work done on the refrigerant in the process of compression causes the temperature of the vapor to increase substantially. The result is a high-pressure, high-temperature, superheated vapor at point 3. The compressor discharge pressure must be high enough to so that the lowest temperature of the refrigerant in the condenser (the **saturation temperature** of the refrigerant) is above the temperature of the surrounding air or seawater used for cooling (the **sink** temperature). It is then possible to remove the superheat and latent heat from the vaporized refrigerant at point 4 in the condenser, so that the refrigerant leaves the condenser at point 4 as a liquid. The cycle repeats, continually "pumping heat" from the **source** (the refrigerated space) to the sink.

Physical laws are not violated: only because the refrigerant in the evaporator is colder than the space does heat flow from the cold space to the even colder refrigerant; and only because of the power consumed by the compressor in doing work on the refrigerant is the refrigerant hot enough to reject heat to the sink in the condenser, and at a pressure high enough to be forced through the expansion device. In fact, in any real refrigeration system, the energy absorbed by the compressor, in the form of power drawn by the motor that drives the compressor, will always be higher than the energy removed from the refrigerated space. Typical marine systems are shown in Figs. 25.2 and 25.3.

25.2 Refrigeration load

Refrigeration load is the amount of heat that must be absorbed and removed from a space to maintain it at the desired temperature. The maximum refrigeration load occurs during **pull down** of temperature during initial startup or when cooling a product initially above the desired temperature. Under these high load conditions the system operates continuously, but once the desired cold temperatures are reached the load is much lower, and the system then **cycles** on and off under automatic control.

Refrigeration equipment is usually rated in its heat-absorbing capacity in units of **tons of refrigeration**. A ton of refrigeration is a unit of thermal energy absorbed per unit time, like horsepower or kilowatts, and should not be confused with the system weight or with the amount of refrigerant within the system. A ton of refrigeration is defined as the amount of heat that must be removed from one short ton of freshwater at 32°F in one day to make one ton of ice. Since the **latent heat of fusion** of ice is 144 Btu/lb_m and a short ton is 2,000 lb_m, a ton of refrigeration equals **12,000 Btu/hr** (3,024 kcal/hr or 3.52 kW).

25.3 Principal system components

Fig. 25.4 is a schematic diagram of a simple vapor-compression refrigeration system, with typical pressures and temperatures listed. The system is effectively divided into two regions by pressure and temperature, the **high-side** and the **low-side**. The dividing points are the compressor, where cold, low-pressure vapor is compressed to a high-pressure, high-temperature vapor, and the expansion device, here (and typical of most systems) an **expansion valve**, where the liquid drops in pressure as it flows through the valve's restricted passage to emerge as a low-pressure, cold mixture of liquid and vapor.

Small refrigeration systems such as domestic refrigerators and window air conditioners work on this simple system. Larger systems are more complex and include additional devices and controls. A typical multi-box refrigeration system is shown in Fig.25.5, and typical components are described below.

(a) Compressor

In the compressor, work is performed on the cold refrigerant vapor to raise its pressure and consequently its temperature above the temperature of the sink.

Compressors may be positive-displacement types or centrifugal. Positive-displacement compressors include reciprocating, scroll, screw, and rotary vane types, with reciprocating compressors being the most common for marine ship's stores refrigeration and for central air-conditioning systems on cargo ships. Scroll compressors are used for small unit refrigerators and air conditioners. Screw compressors are popular with medium-size marine systems. High-pressure or low-pressure centrifugal compressors are common for large-capacity air-conditioning services, as in passenger ships.

Hermetically sealed compressors have both the compressor and the motor mounted inside a welded, gas-tight housing, as in Fig. 25.6. These compressors are used on small appliances such as domestic refrigerators. Since the rotating shaft does not pass through the housing, construction is simplified and inexpensive. Hermetically compressors are not serviced, but are replaced outright upon failure. This type of compressor is used on units up to 5 HP. With **semi-hermetic** compressors (see Fig. 25.7) the compressor and motor are mounted within a common housing, but the housing can be disassembled for compressor repair. This compressor type is used in applications from 5-25 HP. In **open compressors** (see Fig. 25.8) the crankshaft penetrates the housing, and the motor is separately housed. This type is common for larger compressors, about 10-100 HP. Since the compressor crankshaft passes through the housing, the shaft must be sealed against outward leakage of refrigerant or at other times, inward leakage of air. Fig. 25.9 shows a mechanical shaft seal often used for this application.

(b) Evaporator

The evaporator is the heat exchanger through which the cold refrigerant flows to absorb heat from surrounding air. The evaporator may be located in a refrigerated space or in an air-conditioning duct. Inside the evaporator the low-pressure liquid refrigerant boils at low temperature as it absorbs heat from the air in the space or duct. The evaporated refrigerant is then drawn out of the evaporator by the compressor.

To improve heat transfer the evaporator tubes are usually fitted with external fins or other **extended surface** as in Fig. 25.10, and often, in refrigerated stores spaces, a fan is fitted circulate the air.

(c) Condenser

The condenser is the heat exchanger in which the refrigerant rejects the heat absorbed in the evaporator, plus the work of compression, to the sink. A high condenser pressure is necessary for the refrigerant temperature to be higher than the sink, so that heat rejection is possible. Aboard ship the sink for ship's stores and air-conditioning systems is the sea. A water-cooled shell-and-tube condenser is shown in Fig. 25.11. Domestic refrigerators and refrigerated cargo containers use surrounding air as the sink.

(d) Expansion Devices

The expansion device provides a restriction between the high and low sides of the system while at the same time permitting the proper amount of refrigerant to flow. Most marine systems use a **thermostatic expansion valve** or **TXV**, shown in Figs. 25.12 and 25.13.

The TXV controls the flow to ensure slight superheat (about 10°F or 5°C) at the evaporator outlet. To determine superheat, the TXV simultaneously senses both the refrigerant pressure and temperature at the evaporator outlet. With reference to the Fig. 25.12, the temperature is sensed by a thermostatic bulb charged with refrigerant, firmly clamped to the evaporator outlet pipe. The pressure of the refrigerant in the bulb corresponds to the temperature, and is transmitted back to the TXV through the capillary-sensing tube. (The bulb and capillary tube are prominent in Fig. 25.13.) The diagram at top right of Fig. 25.12 shows how the bulb pressure on the diaphragm is balanced by the pressure in the evaporator together with the spring force to operate the valve; the spring force, and therefore the degree of superheat, can be adjusted by a screw on the side of the valve body, as shown at bottom left. In the absence of superheat, the valve throttles the flow of refrigerant into the evaporator, limiting the flow so that the refrigerant is hotter when it reaches the evaporator outlet. If there is too much superheat at the evaporator outlet, the TXV opens, permitting more refrigerant to flow into the evaporator.

The TXV shown in 25.12 is externally equalized, with a pressure-sensing tube connection to the evaporator outlet. However, to simplify installation, in most cases the TXV senses evaporator inlet pressure through a port within the valve body itself. Since the pressure drop in most evaporators is small, internally equalized TXVs are perfectly satisfactory.

For small, simple systems such as domestic refrigerators and window air-conditioning units, a **capillary-expansion tube**, a small-bore tube sized to develop suitable evaporator and condenser pressures at the normal flow rate in the system, can serve as the expansion device. In these systems the small amount of refrigerant in the charge ensures that it vaporizes completely in the evaporator, and a TXV is unnecessary.

25.4 Other system components

Other system components are identified below in the order in which they are encountered in Fig. 25.5, starting at the condenser and following the refrigerant flow counter-clockwise through the cycle. It should be kept in mind that in normal operation, refrigeration systems **cycle** on and off under automatic control to maintain the desired cold temperatures. Some components are also shown in Fig. 25.13.

The **water-regulating valve** controls the flow of cooling water through the condenser. Excess cooling water flow can remove so much heat that the refrigerant pressure on the high side may not be sufficient to force enough refrigerant through the expansion device.

The **receiver** is a surge tank located after the condenser to collect liquid refrigerant during normal off cycles or when the system is to be opened for repair. It may have sight glasses for observing refrigerant level. Marine receivers commonly have two outlets to accommodate the rolling and pitching of the ship.

Relief valves are fitted to the condenser and receiver to protect them from excessive pressure, a possible consequence of an engine room fire. The receiver relief valve discharges to the condenser where the refrigerant is recovered. On some systems the condenser relief valve is fitted with a rupture disk to preclude leakage at normal pressure levels; the disk is designed to rupture at a pressure equal to the set pressure of the relief valve. Small systems are sometimes fitted with fusible plugs designed to melt at high temperature.

A **filter-drier** or **dehydrator** containing a dessicant is fitted in a manually operated bypass. Usually the drier is used only during and after charging to remove any moisture that may enter with the new refrigerant.

A **sight glass** can provide early warning of low refrigerant charge by revealing bubbles in the liquid refrigerant. Some sight glasses are fitted with integral moisture indicators that change color in the presence of water.

The **king solenoid valve** is an electrically operated stop valve that starts the system when a manually operated start button in pressed. When the solenoid is energized the valve opens, permitting liquid refrigerant to flow from the receiver. The system subsequently operates automatically.

An **economizer** or **heat exchanger** enhances the cycle by allowing the high-pressure, warm liquid refrigerant on its way to the TXVs to be cooled by the cold refrigerant vapor leaving the evaporators. The liquid is thereby **subcooled** before reaching the TXVs, increasing its ability to absorb heat in the evaporator and raising the cycle efficiency.

The **box thermostat** is a switch operated by the box temperature, i.e., the temperature in the refrigerated space. When the box temperature rises above the thermostat setting, the thermostat energizes the **box solenoid valve**, allowing refrigerant to flow to the TXV and evaporator for that box. When the box temperature is satisfied, the thermostat de-energizes the box-solenoid valve, stopping refrigerant flow to that box.

An **evaporator pressure regulator (EPR)** is a back-pressure valve installed on each of the higher-temperature boxes of multiple-box systems to force them to operate at higher pressures than the low-temperature boxes. Since the evaporator is at saturation conditions, a higher pressure results in a higher temperature.

An **attemporator** or **desuperheater** is normally found on a system fitted with an economizer, where it serves to limit the superheat of the vapor entering the compressor. It comprises a solenoid valve and a TXV in a branch from the liquid line, and admits liquid refrigerant directly into the vaporized refrigerant.

The **suction-line accumulator** protects the compressor from slugs of liquid refrigerant. By forcing the refrigerant through a tight turn liquid is separated and collects at the bottom of the accumulator. A small metering hole permits the collected liquid to be drawn to the compressor at a low rate. The accumulator is never insulated, and it is common to see the accumulator covered with frost.

The **crankcase-pressure regulator** or **suction-pressure regulator** is a spring-actuated valve that limits the compressor suction pressure during periods of high utilization.

The **low-pressure cutout switch (LPCO)** is the primary control on most systems for starting and stopping the compressor. When all boxes reach their set temperatures, all box solenoid valves close, but the compressor continues to run, lowering the suction line pressure. The LPCO stops the compressor before a vacuum is reached. When box temperatures rise and refrigerant flow is reinitiated, the LPCO senses the rise in suction-line pressure and starts the compressor.

Some large reciprocating compressors are fitted with **unloaders** to allow them to run continuously as space temperatures are satisfied. An unloader is a mechanical device that holds the suction valves open in a cylinder to prevent compression from occurring. As some box temperatures are satisfied a capacity-control system unloads compressor cylinders in succession.

The **high-pressure cutout switch (HPCO)** stops the compressor if the compressor-discharge pressure is too high. Sometimes the HPCO and LPCO are combined in a single housing called a dual-pressure switch.

The **oil separator** removes oil from the hot-vapor refrigerant and returns it to the compressor crankcase.

Refrigeration valves are designed to prevent packing-gland leakage. Two common types are the packless valve and the back-seating valve.

In a **packless valve** (Fig. 25.14) the handwheel and valve stem operate the disk by pressing on a diaphragm that seals the refrigerant in the valve body.

The **back-seating valve** (Fig. 25.15) is similar to a globe valve but has two seats. When the stem is screwed all the way down the valve is front seated or closed. When the valve is opened, the stem is intentionally back seated, which seals the packing gland from the refrigerant, preventing leakage. Commonly, as a second barrier, back-seating valves have a gasketed, pressure-tight cap. Often the cap has handles and a square socket built into the end for serving as a valve handle when inverted.

The **service** (or **charging**) **manifold** (Fig. 25.16) is not part of the system but is an important service accessory, used with a cylinder of refrigerant to charge the system. The manifold has two pressure gages and hoses to connect and read both the high- and low-side pressures simultaneously, and a third hose for connecting to the cylinder.

25.5 Refrigerants

A good refrigerant must evaporate and condense at temperatures required for a particular service, and at reasonable suction and discharge pressures. It should be non-toxic, chemically inert, non-reactive with air, moisture, oils, and system materials, be non-contaminating for food, and have high latent heat of vaporization. A low condensing pressure is desirable.

Fluorocarbons (fluorinated hydrocarbons) are by far the most prevalent refrigerants in use today. These are broken down into the **ChloroFluoroCarbons (CFCs)**, the **HydroChloroFluoroCarbons (HCFCs)**, and the **HydroFluoroCarbons (HFCs)**. Under international law, the CFCs and HCFCs are being prohibited from new manufacture because they deplete the ozone layer.

CFCs are considered the most harmful to the environment and have been banned from manufacture, although they remain in use in older systems. Some examples of CFCs are:

Trichloromonofluoromethane	CCl_3F	R-11	74.9°F (23.8°C) boiling point at 1.0 atm.
Dichlorodifluoromethane	C_2Cl_2F	R-12	-21.6°F (-29.8°C) boiling point at 1.0 atm.
Monochlorotrifluoromethane	C_3ClF	R-13	-114.6°F (-81.4°C) boiling point at 1.0 atm.

R-11 was a low-pressure refrigerant used in large centrifugal air conditioning systems. R-12 was the common general-purpose refrigerant used in most applications. R-13 was a low-temperature refrigerant used for very low-temperature applications.

The HCFCs are also considered to be hazardous to the environment, although less so than the CFCs because they begin to chemically break down before reaching the stratosphere. Consequently, they are are still legally manufactured, but a manufacturing ban will go into effect before 2010. Examples of HCFCs include R-22 (atmospheric boiling temperature of -41.3°F/-40.7°C) and R-123 (atmospheric boiling temperature of 83.7°F/28.7°C).

The HFC refrigerants contain no chlorine atoms and are presently considered to be environmentally friendly. HFCs include R-134a (atmospheric boiling temperature of -15.7°F/-26.5°C), R-124 (atmospheric boiling of 10.4°F/-12°C), and some of a new 400 series. It was initially hoped that R-134a would be a "drop-in" replacement for R-12 and R-22, but it has shortcomings. In addition to HFCs, ammonia, with an atmospheric boiling point of -28°F (-33.3°C), has come back into use. Good replacements for the old standards continue to be sought.

Recovery, recycling, and **reclamation** are important goals in servicing refrigeration systems. Recovery is the removal and safe storage of refrigerant from a system. Fig. 25.17 shows a recovery unit. Recycling is the filtering, drying, oil separation, and acidity neutralization of a refrigerant for reuse in the same system from which it was recovered (recycled refrigerant should never be used in any system other than that from which it came). Reclamation is the reprocessing of recovered refrigerant at a manufacturer's facility for general resale and reuse.

Recovery, recycling, and reclamation of CFC refrigerants is encouraged, and commerce in these refrigerants is not taxed. The recovery goal is 95% of a full charge. Release of these refrigerants to the atmosphere must be avoided, and release of quantities exceeding minimal levels is punishable by very heavy fines and imprisonment. Refrigeration technicians must be certified. Only refrigerant manufacturers are permitted to destroy refrigerant.

25.6 Operation of refrigeration systems

(a) Starting and stopping

Refrigerant systems generally operate automatically once started. With reference to Fig. 25.5, depressing the start button opens the king solenoid valve, allowing liquid refrigerant to flow to the boxes. The box thermostat determines whether refrigerant will be admitted to a particular box: on a rise of temperature in the box, the thermostat closes, energizing the box solenoid to open the box solenoid valve. Refrigerant is then flows through the TXV, which automatically throttles the flow to maintain a constant amount of superheat at the evaporator outlet. As the refrigerant flows through the evaporator, the compressor suction pressure begins to rise. This suction pressure is sensed by the LPCO switch, which closes when its high-pressure (cut-in) setting is reached, starting the compressor.

The compressor unloader starts the compressor in the unloaded condition, and based on suction pressure, it loads the number of cylinders required to match the refrigeration load.

As each box temperature reaches its intended cold temperature, the box thermostat closes the box solenoid valve. If all boxes are cold and all box solenoids are closed, or if the stop switch de-energizes the king solenoid, the system goes into a **pump-down cycle**. The compressor continues running, unloading cylinders, until the suction pressure is low enough to actuate the LPCO, stopping the compressor.

(b) Defrosting

Defrosting removes ice from the air side of the evaporator. The time between defrost cycles depends on the type of evaporator, the installation, the operating temperature, and the defrost method. Large bare-tube evaporators may require defrosting once or twice a month, while small, fin-tube, forced-convection evaporators may require defrosting every few hours.

Natural defrosting, where the evaporator is shut off and the ice melts, may be used in spaces at temperatures of 37-40°F (2-4°C). In colder spaces, supplementary heat from electric resistance heaters or from hot compressor-discharge refrigerant re-routed to the evaporator (**hot gas defrosting**) may be used. Another defrost method is to spray a hot-brine solution onto the non-operating evaporator. Defrost operation may be initiated manually or by timer.

(c) Charging a system

When the system is first installed or after extensive repairs, it must be filled with refrigerant. Air and moisture must first be **evacuated** from the system with a vacuum pump. The correct amount of refrigerant, called the **charge**, is installed by weighing the refrigerant cylinder or bottle during the charging process.

In one charging method, shown in Fig. 25.16, liquid refrigerant is drawn into the system by the compressor through the liquid-charging connection, using the service manifold. Alternatively, vapor refrigerant can be added through the vapor charging connection at the compressor suction. While charging, the compressor is operated by the LPCO switch to draw the refrigerant into the system. The cylinder may have two valves, red for vapor and blue for liquid (fitted with a dip tube to the bottom of the cylinder). Bottles with a single valve are inverted to deliver liquid. During and after charging, the refrigerant is directed through the dehydrator until it is certain that the system is moisture-free.

(d) Adding oil to the compressor

The compressor oil level is indicated by a gage glass in the crankcase, and should be maintained one-half to two-thirds full when the compressor is running. Oil and refrigerants (other than ammonia) are miscible, and the oil level of a secured compressor may appear to be high, but will drop after the compressor starts as the refrigerant boils out of solution.

The oil used must be compatible with the refrigerant. Oil is added using a hand-operated oil pump whose suction is dipped into a can of clean, compatible refrigeration oil, as illustrated in Fig. 25.18. The compressor oil filter inlet is temporarily disconnected, and the hand-pump discharge is connected to the filter. To avoid air infiltration, the pump connection is first loosely made and the hand pump operated until oil begins to issue from the connection. The connection is then tightened and the oil is pumped into the crankcase until the proper level is reached. Partially used oil containers must be tightly sealed to exclude moisture.

(e) Leak detection

The easiest means of leak detection is by looking for oil. Since most refrigerants are miscible with oil (ammonia is an exception) lubricating oil also circulates throughout the system. If the refrigerant leaks, then the oil also leaks, leaving a telltale residue. Phosphorescent tracers and black lights can help to find the residue.

A traditional leak detection device used with refrigerants other than ammonia is the **halide torch**, shown in

Fig. 25.19. The torch uses a portable fuel tank and its flame is directed against a copper plate. A sampling hose is connected to the torch handle. The flow of propane produces a venturi effect, drawing air through the hose to the flame. In the presence of halogen gases the light-blue flame changes color to bluish-green or purple. When burned, halogen gases produce phosgene gas, a deadly nerve gas. Halide torches should only be used in well-vented areas.

Electronic sniffers are also used. These devices are specific for each refrigerant.

CHAPTER 26
HEATING, VENTILATION, AND AIR CONDITIONING

26.1 Air conditioning

When people think of air conditioning they think of cooling for comfort in hot weather. But air conditioning actually includes the control of temperature, humidity, air purity, and air movement in either hot or cold weather, for comfort and to protect equipment. The acronym **HVAC** is in common use for heating, ventilation, and air conditioning.

Comfort is the sensation derived not only from air temperature, but also from humidity levels and air movement. In the summer people wear short-sleeved shirts to allow perspiration to evaporate, providing a natural, evaporative-cooling effect. It is common to use fans to accelerate evaporative cooling. Mechanical air conditioning is used to reduce space temperature and humidity levels for comfort, and to avoid the growth of mold.

In the winter people wear long-sleeved, heavy clothing to insulate the body against heat loss and for protection from drafts. Space heating improves comfort but reduces relative humidity. Low humidity is undesirable because it promotes evaporative cooling from perspiration and dries the mucous membranes.

Therefore, properly designed and operated air-conditioning systems must maintain acceptable humidity levels as well as comfortable temperature levels. Generally, 30-70% relative humidity are considered the healthy limits, with comfort goals of temperatures between 68-72°F (20-22°C) and levels of 40-50% relative humidity in cold weather, and 78-82°F (26-28°C) with 50-60% relative humidity in warm weather.

26.2 Relative humidity

Relative humidity is expressed on a percentage basis: 0% relative humidity represents perfectly dry air while 100% relative humidity is completely saturated air, or air in which no further evaporation can occur. Relative humidity, rather than absolute humidity, is the factor that affects comfort, since relative humidity affects the rate of evaporation of perspiration.

The **dew point** is the temperature at which water vapor in the air begins to condense. At the dew point the relative humidity is 100%. Cooling to temperatures below the dew point results in **dehumidification** of the air as the water vapor condenses and precipitates.

The **dry-bulb temperature** is the normal ambient temperature as read from a common thermometer. The **wet-bulb temperature** is the temperature read from a thermometer whose bulb is wrapped in a continuously wetted by a wick. As the water evaporates from the wick, the wick temperature drops below ambient. The rate at which the water evaporates from the wick is related to the amount of moisture contained in the surrounding air, i.e., the humidity. From the dry-bulb and wet-bulb temperatures, the humidity of the air can be found.

A **sling psychrometer** (see Fig. 26.1) is a set of wet and dry-bulb thermometers mounted side-by-side. The psychrometer is slung around a swivel joint to provide a flow of air over the thermometers, aiding evaporation of moisture from the wick of the wet bulb. The dry-bulb and wet-bulb temperatures are correlated to yield the relative humidity. An electronic **thermohygrometer** can provide temperature and relative humidity directly on a digital display.

26.3 Ventilation and exhaust; fans

Ventilation of a space replaces stale air with fresh air. Proper ventilation is measured by the number of **air changes** required to reduce odors, airborne particles, and accumulated carbon dioxide from respiration, and the removal of flammable and toxic vapors. Ventilation requires an air supply and outlet from each space. On the supply side are air handlers, consisting of a fan, motor, and housing. **Exhaust fans** are always fitted to spaces like heads, galleys, engine rooms, pump rooms, and purifier rooms to pull air from the space and direct the exhaust to the weather directly or via ductwork.

Common types of fans include **centrifugal, vane-axial,** and **tube-axial fans** (see Fig. 26.2). Centrifugal fans can develop a higher pressure than axial fans and are suitable for ducted distribution systems, as in Fig. 26.3. The vane-axial and tube-axial fans use rotating, vaned wheels mounted in cylindrical housings, which can be incorporated into circular ductwork, often with the motor inside the duct as shown. The vane-axial fan is similar to the tube-axial fan, but has fixed vanes after the rotor to straighten and improve the air flow. Axial-flow fans are limited in the amount of pressure that they can develop and are therefore likely to be found in simple duct systems, including dedicated air supply systems and direct exhaust systems.

Propeller fans are axial-flow fans generally used independently of ducted systems. A common application is the **unit heater** shown in Fig. 26.4, which combines a fan and heating-coil unit. Unit heaters are usually used for heating unoccupied spaces, such as shaft alleys and steering-gear rooms. The fan is usually switched on and off by a thermostat that senses space temperature, with a continuous supply of steam or hot water to the heater.

Fan **capacity control** is achieved with dampers or by varying fan speed. Dampers at the fan intake are preferred over discharge dampers because they reduce energy consumption, improving overall efficiency. Some shipboard fans use multiple-speed motors.

Fans may be direct-driven as in these illustrations, but belt-driven fans are also used. Belt drives offer the ability to match fan speed, capacity, and pressure to the application, and permit a smaller, less costly, high-rpm motor to be fitted.

26.4 HVAC systems

A typical system is shown in Fig. 26.3. The ducted supply path from the air handler to the space is evident. Exhaust air exits the space through the door louver into an adjacent passageway, from which it is drawn to the air handler through the return duct. Some of the exhaust air is discharged to the weather but the rest is recirculated, after mixing with fresh ambient air drawn in from the weather. The system must be designed to ensure that the exhaust air is discharged sufficiently far from the ambient air weather intake to avoid drawing it back into the system. A typical central-station system is shown schematically in Fig. 26.5.

Dampers are adjustable sheet-metal baffles located in the ductwork to apportion the flow of air in the system.

A **plenum** is a void space in the air handler housing. Air filters are installed in the plenum, at the inlet of the fan. Larger air handlers may have plenums large enough for a person to enter in order the change or clean the air filters. The air filters are usually installed on the inlet side of the fan in order to reduce dust and other particles that can adhere to the fan blades and imbalance the fan impeller, or can foul the heat-exchanger coils.

A heat-exchanger coil in a ducted system usually comprises a set of heated or cooled tubes with inlet and outlet headers for circulating the hot or cold fluid. On their air sides the tubes are usually fitted with **fins** or other **extended surface** to improve heat transfer. The **cooling coil** cools and dehumidifies the supply air. By chilling humid air to about 50°F (10°C), well below the dew point in uncomfortably humid conditions, enough moisture will condense and precipitate, so that after reheating to a comfortable temperature the remaining humidity will be within desired levels. The tubes of the cooling coil are usually circulated with chilled water.

In hot weather, the **heating coil** or **reheat coil** is used to reheat air that has been chilled below its dew point for dehumidification. In cold weather the heating coil is used together with the **pre-heater** coil to warm the spaces to comfortable temperatures. The tubes of the heating coils are usually heated by steam or hot-water.

A **humidifier** may be fitted for use in cold weather to raise humidity levels. A humidifier may use direct steam injection, water spray, or water wicked onto a membrane located in the air stream.

26.5 Compressors, chillers, and condensers

Air-conditioning systems up to about 100 ton capacity, suitable for cargo ships and smaller passenger and naval vessels, usually use reciprocating compressors and are similar in appearance and operation to the ship's-service refrigeration system. However, the air-conditioning system is likely to be an **indirect-expansion** system as in Fig. 26.6, where the refrigerant cools water (usually mixed with anti-freeze) to about 40°F (about 5°C), in a **chiller/evaporator**. See also Fig. 25.2. The refrigerant evaporates on the shell sides of the tubes in the chiller/evaporator, chilling the water in the tubes. The chilled water is then pumped to remote coiling coils located in ducts or in **fan rooms** near the cooled spaces. Larger shipboard systems with loads over 100 tons, such as those for passenger ships or naval vessels, may use **centrifugal-compressors** as shown in Fig. 26.7, or rotary-screw compressors.

CHAPTER 27
HULL MACHINERY

27.1 Introduction and scope

This chapter will describe some of the hull machinery commonly found aboard ship, including steering gear and such deck machinery as windlasses and winches. Other hull machinery and equipment, including stabilizers, elevators, thrusters, and inert-gas generators, is not included.

Steam-driven hull machinery was once common but is generally obsolete. Modern steering gear is invariably electrohydraulically driven and almost all deck machinery is driven by electric or hydraulic motors. Electric motors are discussed in Chapter 24; basic principles of hydraulic machinery are outlined in the following section.

27.2 Hydraulic machinery

In its simplest form a hydraulic system might comprise a pump, piping with valves and other fittings, a reservoir, and an **actuator**, as in Fig. 27.1. The pump is usually driven by an electric motor but in some applications a steam turbine or diesel engine might be used. The **hydraulic fluid**, which is a carefully selected oil, is discharged by the pump under sufficient pressure to operate the actuator. The actuator may be a cylinder with a piston or a rotating motor. The oil exhausted by the actuator may return directly to the pump in a closed loop, or to the reservoir. In unitized systems a single motor-and-pump, called a **power unit**, is dedicated to serve a single actuator. In **central systems** a group of pumps supplies oil continuously to a pipe main, from which a number of actuators can be supplied.

The pumps are usually positive-displacement pumps. In a simple system, a fixed-delivery pump of the screw, gear, or vane type (see Chapter 20) runs continuously, but until called upon to operate the actuator, the oil discharged by the pump passes through an unloading valve back to the pump suction. When actuator movement is required, a valve is opened to pass the oil to the actuator, while the unloading valve simultaneously closes.

Variable/reversible-delivery pumps are also used. These are mechanically complex pumps that have the ability to provide continuously variable flow, from maximum in one direction to a maximum in the opposite direction, even while the pump is being driven in one direction at constant speed. The pump runs in neutral, with no flow, except when called upon to move the actuator.

Several types of hydraulic motors are in use, but one example is the **vane motor**, shown in Fig. 27.2.

A type of valve common in hydraulic circuits is the **sliding-spool valve**, illustrated in Fig. 27.3. These valves are readily operated by a lever attached to the spindle, but can also be arranged for operation by an electric solenoid, which facilitates remote control and automatic operation.

Hydraulic systems are compact yet capable of very high forces, are very reliable, and are far more resistant to weather and moisture damage than electric systems. They are quickly started, offer high torque on demand, are easily controlled, and lend themselves well to automation.

27.3 Steering gear arrangements

The steering gear must position and hold the rudder against the force of the sea. The force acting on a rudder is proportional to the rudder area, the rudder angle, and the square of the speed of the ship. The resulting force is large, and rules generally require that the major parts of the steering gear, including the **rudderstock**, the tiller and the actuators, be of steel. The weight of the rudder and rudderstock is supported in a **rudder carrier bearing** usually built into the base of the steering gear.

The three types of steering gear in common use today are all electrohydraulic: the hydraulic oil pressure is normally supplied by one of a pair of duplicate power units. Both power units are used during maneuvering periods. The types of steering gear differ in the arrangement of the actuators.

Ram-type steering gear: A typical arrangement for a four-cylinder ram-type steering gear is shown in Fig. 27.4. Normally only one power unit is in service at a time. When a rudder movement is required oil from the pump, typically at 1500 to 2500 psi (100 to 175 bars), passes through control valves to two diametrically opposed cylinders (while exhausting from the other two cylinders) to move the rams in opposite directions. The rams act through the trunnion, trunnion block, and yoke (which serves as the tiller) to turn the rudderstock. Because of the arrangement of the trunnion block in the slot of the yoke, this type of steering gear is called **Rapson-slide steering gear**.

An order for rudder movement is transmitted from the autopilot or steering wheel at the bridge to the **differential assembly**, which operates the pump controls or valves to initiate the flow of oil to the cylinders. The helix shaft and **follow-up assembly** trace the movement of the rudder, and act through the differential assembly to stop the oil flow when the rudder is in the ordered position. In emergencies the **trick wheel** allows local control. The hand pump can be used to position the rudder in the event of a complete electric failure.

Rotary-vane steering gear is shown in Figs. 27.5 and 27.6. The actuator consists of a housing or stator, which is fixed to the foundation and contains two vane cavities, and a rotor with vanes attached, which acts as a tiller. When a rudder movement is required oil from the pump is directed by the control valves to the vane cavity on one side or the other of each of the rotary vanes (while exhausting from the vane cavity on the other side). The high-pressure oil acts on the side area of the vanes to turn the rudder. Rotary-vane actuators are more compact than ram types, and are suitable for lower operating pressures, generally 850 to 1450 psi (60 to 100 bars). Control of the rudder is accomplished as described above for ram-type steering gear.

Clevis-mounted steering gear is used extensively on smaller vessels. One example is shown in Fig. 27.7, for a twin rudder vessel with canted rudders. Clevis-mounted units utilize hydraulic cylinders, one end of each of which is pinned to ship's structure, while the piston rod end is attached to the tiller, also by a pinned connection. The pumps are not shown Fig. 27.7. Control of the rudders is accomplished as described above for ram-type steering gear.

27.4 International agreement on steering gear capabilities and operation

Requirements for steering gear capabilities are set by governmental regulatory authorities and by classification societies, but generally in accord with the International Convention for the Safety of Life at Sea of 1974 (SOLAS 74), which has been revised and amended, but remains in effect. In general, the requirements call for increasingly higher levels of redundancy and capability of power units, actuators, piping, electric supply, and control systems as the risk increases. In most vessels, for example, power units, power supplies, and control systems must be duplicated, while tankers and gas carriers are additionally required to have duplicate rudder actuators.

Requirements for the operation of steering gear are also set out in SOLAS, generally as follows:

- Where navigation requires special care both power units shall be in use.
- Within 12 hours prior to departure, the steering gear and linkage shall be inspected and tested

over the full movement of the rudder. For ships engaged in short voyages these inspections and tests shall be conducted weekly.

- Operating instructions with a block diagram showing emergency procedures shall be permanently displayed on the bridge and in the steering gear compartment. Drills shall be conducted at least every three months, and shall include direct control from the steering gear location, and the use of alternative power supplies.
- Checks, tests, and drills shall be recorded in the logbook.

27.5 Windlasses, winches, and capstans

Windlasses are used to raise and lower ships' anchors. **Winches** and **capstans** are used to handle wire or rope lines. These machines have some features in common: a prime mover, which may be a hydraulic or electric motor, a speed-reducing gear, and a drum or head to haul on the chain or line.

Because of the fire hazard on tankers and gas carriers deck machinery is generally driven by hydraulic motors, usually supplied from a central electrohydraulic system. On ships of other types the deck machinery may be driven by hydraulic or electric motors.

A windlass is shown in Fig. 27.8. Because windlasses are used for hoisting (or "**heaving**") anchor chain, the drum takes the form of a sprocket called the **wildcat**, which is shaped to grasp the links of the chain, as in Figs. 27.8 and 27.10, in alternating **whelps** and pockets cast into the sprocket. This example is a duplex windlass, used to hoist both anchors. The chain is not stored on the windlass but passed through a **chain pipe** in the main deck to a **chain locker** below.

In Fig. 27.8 the electric motor is in the foreground, driving into the gear case. The output shaft from the gears passes right and left through the hubs of the wildcats. For each wildcat, a lever-operated **clutch** engages the wildcat to the shaft. The large hand wheels operate the **brakes**, one for each wildcat. The windlass can hoist both anchors simultaneously at low speed, or each anchor separately at higher speed. This windlass is also fitted with warping heads (see below). These heads are not clutched, and turn whenever the motor is driving the windlass, but are used only when both of the wildcats are disengaged and locked by the brakes.

In operation, with the clutch disengaged, the anchor will fall under its own weight as the chain turns the wildcat, until the brake is used to control the fall. To hoist the anchor, or to lower the anchor under power, the clutch is engaged. The windlass is not used to secure the anchor at sea: once the anchor is heaved up and firmly seated at the foot of the **hawsepipe**, a **chain stopper** is used to lock the chain in place.

Winches are used to handle wire or fiber rope lines. Fig. 27.9 shows a typical hydraulic **mooring winch**, driven by the motor in the foreground, via the reduction gears. Similar units are driven by electric motors. This winch has a split drum, with a clutch and hand-operated band brake. The winch in Fig. 27.10 also has a split drum. **Split drums** prolong the lives of the rope lines, by allowing the length of line not in use to remain loosely spooled on the drum on one side of the dividing flange, while the active part of the line is worked by the other part of the drum. The tensioned part of the line is thereby kept from working in among the coils of stored line, which could then become frayed and damaged.

Fig. 27.10 shows a winch on the left and a windlass on the right, but both are driven by the single hydraulic motor at the center. When driving the windlass, the clutch for the winch drum is opened and the drum is secured by the winch brake. The output shaft of the winch reduction gear, which extends to the right of the winch gear for the purpose, can then be clutched to the pinion in the windlass gear case. The same motor that has sufficient torque to drive the winch through a single-stage gear train can thus provide the greater torque needed to hoist the anchor, because of the mechanical advantage of the second stage gear train.

Mooring winches are normally operated under manual control to set the tension in the mooring line that is needed to hold the ship in place at the berth, at which point the brake is set tightly enough to hold the ship, but loosely enough to slip should line tension rise as the ship is loaded or discharged or as the tide changes. Therefore, while a ship is moored to a berth, a mate will periodically adjust the lines. Automatically tensioned mooring winches (also called self-tensioned or constant-tension winches) sense the tension in the mooring line and control the winch automatically, paying line out as tension on the line is sensed to rise, and taking line in when slack.

Capstans and **warping heads** (also called gypsy heads) are similar to each other in construction, purpose, and use, but differently oriented: capstans are vertical while warping heads are horizontal. While warping heads are often fitted to the extended shafts of windlasses and winches as noted and illustrated above, capstans are most often driven by independent motors. Fig. 27.11 shows a capstan driven by an electric motor, with the motor and reduction gear mounted to the deck below.

Capstans and warping heads are used to assist in the manual handling of fiber rope lines: the person handling the line wraps it once or twice around the rotating head, and then, standing behind the head, applies sufficient tension to enable the head to grasp the line by friction. The operator pulls the line hand-over-hand to maintain the tension, coiling the line on deck or passing it to a bin or storage spool below through an open hatch.

27.6 Arrangement of deck machinery

Fig. 27.12 shows the foredeck of a tanker. Two windlasses are visible, with the anchor chains led to the hawsepipes over the chain stoppers. Also in view are four duplex, split-drum mooring winches. All of these windlasses and winches are hydraulically driven, with the windlasses driven through a clutch from the adjacent winch as described above. Numerous fairleads and chocks are available to enable mooring lines to be passed to bitts on a pier. Additional winches are likely to be located along the main deck of the ship, with another two or three winches grouped together at the stern.

Chapter 28
SAFETY EQUIPMENT

28.1 Introduction

Seafaring has always been a dangerous undertaking. The sea is a hostile and unforgiving environment, putting those who venture onto it at peril. Great tragedies of the twentieth century led to significant advances in ship safety and survivability. The sinking of the *Titanic* triggered new regulations for lifeboats. Passenger vessel fires in the mid-1900s led to new regulations for ship construction materials. But despite significant technological advances and the broadening scope of ship safety regulations, accidents still occur.

The great majority of accidents at sea are the direct result of human error. Accidents usually involve collision or grounding, and often result in flooding or fire, two of the greatest perils of the sea. The professional competence of the crew is a major factor in ensuring marine safety. Adequate crew training and proper maintenance and testing of safety equipment are essential to the safe operation of marine vessels.

Fire prevention is critical to ship safety and survivability. Prevention begins with training. Fires can spread rapidly and are best dealt with as early as possible, so early detection is very important. Once a fire is detected fire prevention gives way to firefighting. Training is also critical in firefighting, as are ship construction, arrangement, and extinguishing systems.

If flooding or fire overwhelms the vessel and its crew, the master will give an order to **abandon ship**. Here too, crew training and competence are critical to survival.

Merchant ship safety is regulated by the U.S. Coast Guard (USCG) for US-flag ships and for international shipping in US territorial waters. The International Maritime Organization (IMO) establishes safety standards for international shipping, including US-flag ships on international voyages. IMO adopted the latest Safety of Life at Sea convention (**SOLAS**) in 1974, to which most seagoing ships are subject. SOLAS specifically addresses lifesaving equipment and fire safety issues. USCG has adopted many of the SOLAS requirements as their standard of safety. Classification societies, such as the American Bureau of Shipping (ABS), certify ships according to safety standards, and generally use the SOLAS requirements as their benchmark. Unlike the USCG, whose function is to safeguard the public, property, and the environment, the classification societies' function is to safeguard the interests of the ship owner and insurance underwriter. Note however that the USCG uses much of the classification society standards to accomplish its safety mission.

28.2 Lifesaving equipment

Lifeboats are **primary lifesaving equipment**, which provide out-of-water support for crew and passengers forced to abandon ship. They are normally required to be of sufficient size and quantity to accommodate the total number of people on board, usually with half the boats carried on each side of the vessel. Lifeboats vary considerably by type, size, and passenger capacity. **Open lifeboats** and semi-enclosed lifeboats are used in limited applications such as harbor and river craft. **Totally enclosed lifeboats**, as shown in Fig. 28.1, are generally required for all ships over 85 m (279 ft) in ocean service, and offer better protection against the weather and sea. Totally enclosed lifeboats, with their extra top buoyancy, are usually self-righting. Cruise ship **tenders**, used to ferry passengers ashore where berthing for the ship at a pier is unavailable, do double duty as lifeboats. Fig. 28.2 shows a tender lowered to its boarding position. Some lifeboats, particularly those for tankers and offshore (oil) production platforms, are fitted with sprinkler systems to envelop the boat with a mist of seawater if sailing through burning oil. These sprinkler systems are required to resist flames for an eight-minute period, and the boats are fitted with their own air supply for this interval.

Lifeboats have different means of propulsion. Enclosed lifeboats are fitted with small diesel engines, but may also be oar-propelled if the engine is disabled. Motor lifeboats must have sufficient fuel for 24-hour continuous operation. They must be capable of ahead speed of 6 knots (7 mph) in smooth water with a full load.

Lifeboats are usually launched from **davits. Gravity davits**, as in Figs. 28.1 and 28.2, require only a manual release to initiate a launch. Davits must be arranged so that the boat need not first be lifted from a boat **cradle** before the boat is lowered. Typically, a system of wire ropes (the **lifeboat falls**) and pulleys are used to lower the lifeboat to the **boat deck** for embarkation, and then to the water. Most lifeboats are required to be fitted with mechanical **disengaging gear** that allows both ends of the lifeboat to be released from the falls simultaneously by one person once the boat is in the water. Davit-launched lifeboats must be capable of being launched fully loaded when the ship has up to 10 degrees of trim and a list of up to 15 degrees either way. When lowered for **boat drills** or maintenance, lifeboats are raised by a winch, using the falls.

Free-fall lifeboats are often fitted to ships and offshore rigs. Free-fall boats are dropped from as high as 40 m (over 130 ft). A free-fall lifeboat on a ship is typically launched over the stern in an articulated slide, as shown in Fig. 28.3. The boats are totally enclosed and self-righting. They are fitted with rear facing seats and safety belts to minimize the shock and stress of the fall. The **coxswain** in charge of the boat has 360-degree vision. Launching may be initiated mechanically or by a float-free arrangement. A free-fall lifeboat may be supplemented with davit-launched liferafts. When launched for training or testing, a free-fall boat must be retrieved with a crane and returned to its launching position.

Liferafts and other **inflatables** such as **inflatable buoyant apparatus** (IBAs) are usually **secondary lifesaving appliances** for ships. They may constitute primary lifesaving gear in special circumstances such as for vessels in non-ocean service. Liferafts vary in size, shape, and function. Shipboard liferafts may be fitted for as few as six persons, or as many as 100 persons. Generally, the higher-capacity liferafts are employed for high-density passenger vessels on short international voyages. **Lifefloats** are rigid, buoyant, non-inflatable platforms serving a similar function as liferafts. Lifefloats are less costly than liferafts, but liferafts make much more efficient use of valuable deck space on the ship.

Liferafts and IBAs are packed into fiberglass containers. The containers are normally cylindrical in shape, split into two parts down the length of the cylinder. The liferaft is inflated by CO_2 from storage canisters packed with the raft inside the container. Liferafts may be open or covered. They may be self-righting. A typical 25-person liferaft weighs about 185 kg (over 400 lb). IBAs are open, with a buoyancy ring at the perimeter and a circular platform within the ring. A launching rack with three IBA canisters is shown in Fig. 28.4.

Liferafts may be launched from davits, cradles, or free-fall racks. Davit-launched rafts are usually launched from a single davit. They are inflated on board, boarded, and then lowered to the water. Cradle-launched liferafts are either thrown overboard manually or are fitted with **float-free devices** that allow the liferafts to float to the surface automatically if the ship sinks. The most common float-free device is a hydrostatic release that releases the liferaft container lashings at a depth of 2 to 4 m (about 6 to 13 ft). The buoyancy of the raft brings it to the surface where it remains attached to the vessel by a **painter** (a length of rope) until it is cut free by the crew. **Free-fall racks** allow the rafts to be deployed mechanically, and may also be fitted with float-free devices.

Rescue boats are small, lightweight boats designed to rescue people in distress and to marshal and tow survival craft (liferafts and IBAs). They are designed to be launched in minutes. They must remain stable when recovering an unconscious person from the water over the side of the boat. Rescue boats come in various shapes and sizes and are normally davit-launched. A typical example is a semi-rigid boat with a fiberglass hull and inflated rubber buoyancy chambers for extra stability, as shown in Fig. 28.4.

Inflatable escape slides are similar to aircraft escape slides. They are typically found on large passenger ferries and on cruise ships. They are capable of evacuating 360 passengers in 30 minutes. Single slides are used from decks up to 6 m (20 ft) above the water. Dual (tandem) slides are used from decks up to 15 m (49 ft) above the water. The escape slide is part of a four-component system:

- a steel stowage box containing the deflated slide and bottles of compressed nitrogen for inflation
- the slide itself
- an embarkation platform at the foot of the slide
- liferafts or IBAs

Inflation of the slide normally takes 2-5 minutes. The platform serves as a buffer zone until the first liferafts are deployed, inflated, and then marshaled into position by a rescue boat. **Escape chutes** serve essentially the same purpose as escape slides. They consist of enclosed fabric (usually nylon) tubes. As shown in Fig. 28.5, passengers enter the tube at an upper deck and drop down to an embarkation platform. Their descent is slowed by friction and, in some designs, by baffles.

Life preservers, also referred to as **personal floatation devices** (PFDs), come in a number of shapes, sizes, and designs. They may be of the solid buoyancy type (e.g. closed cell foam) or may be inflatable. Inflation is either oral, by a CO_2 cartridge, or a combination. All of the inflatables must be capable of being topped off orally. Lifejackets are typically fitted with ball-type whistles attached by a three-foot lanyard, a water-activated light or chemical light stick, and reflective material.

Ring life buoys are fitted at the perimeter of the ship's weather deck to be thrown rapidly to a person overboard. They are connected to the ship by a painter.

Survival suits, also called exposure or immersion suits, are protective coveralls that reduce the body-heat loss of a person in cold water. They are typically designed to keep the body temperature of a person in water at 0 °C (32 °F) from falling below 35 °C (95 °F) for six hours. These suits are often required for vessels fitted with open lifeboats.

28.3 Communication systems

All ships are fitted with **general alarm** systems, which are used to alert and summon the crew to their **fire stations** or **boat stations**. Passenger vessels are required to have public address systems. Portable very high frequency (VHF) radios, commonly referred to as **walkie-talkies**, are required for emergency crew communications. Non-sparking walkie-talkies are required where an explosion hazard exists, as on tankers.

28.4 Fire prevention and firefighting equipment

The risk of fire is high aboard ship. The three elements necessary for a fire, oxygen, fuel, and an ignition source, are all normally present and in close proximity. Leaking fuel oil, hot lubricating oil, paint, and solvents can be ignited by overheated equipment, electric short circuits, hot turbochargers and exhaust piping, or other ignition sources. Potentially hazardous locations in engine rooms, flammable-liquid storage rooms, and galleys are usually fitted with detectors of a type suited to the particular threat. For example, machinery spaces are fitted with **heat detectors** because of the major threat of petroleum fires. Passenger vessels are typically fitted with heat detectors and **smoke detectors**. Depending on the cargo carried, cargo spaces may be fitted with **flame detectors**.

Firefighting systems aboard ships include fire mains, fixed fire-extinguishing systems, and portable firefighting equipment. All ships are fitted with **fire-mains** to enable seawater to be pumped to **fire stations** in strategic locations throughout the ship. Fire mains are used for most fires, although the fact that salt water is an electric conductor makes it unsuitable for fighting electrical fires. A major concern when seawater is used for fire fighting is that if an excessive quantity accumulates on board it can destabilize or sink a ship.

Every ship of 500 gross tons or more is required to have two independently driven and physically separated fire pumps. Sanitary, ballast, bilge, or general service pumps may serve as fire pumps provided they are not normally used for pumping oil. In general, the pumps are located in separate compartments so that a fire in one compartment will not render both pumps inoperable. Therefore, on most ships, one fire pump is located outside of the main machinery space. In addition, most ships are required to have an emergency fire pump. The emergency fire pump and its power source must be located outside of the main machinery space.

Fire stations, as shown in Fig. 28.6, consist of a **hydrant** (valve), **fire hose** and hose rack, a **spanner** wrench (for attaching the hose), and a hose **nozzle**. There must be enough hydrants, and they must be so located, that any part of the vessel accessible to the passengers or crew while the vessel is underway, including the cargo holds of dry-cargo ships (but not the main machinery spaces) may be reached by two streams of water from separate outlets. At least one of the streams must be from a single length of hose.

Fire hose is normally 50-feet long, 2-1/2-inch diameter. Interior hydrants may substitute a **siamese connection** for 1-1/2-inch hose, with a single length of 75-foot hose. Each fire hose on each hydrant is fitted with a combination solid stream and water spray fire hose nozzle. Machinery space fire stations also have **low-velocity water-spray applicators** or wands (see Fig. 28.6) used to cool the space and allow firefighters access to the source of the fire.

In contrast to fire hoses, **fixed firefighting systems** include fixed piping for distributing the firefighting medium. Examples include the **sprinkler systems** that are fitted on some ships to meet special needs, as in passenger cabins and in the vehicle decks on ferries. Special cargoes may also warrant sprinkler systems in the cargo holds. **Water mist systems** have been developed for applications requiring little water and short spray duration.

Fixed firefighting systems are also required in engine rooms of almost all ships, and in the pump rooms of oil tankers. These are generally **total-flooding fire-suppression** systems using carbon dioxide (CO_2). CO_2 is colorless and odorless, normally harmless to equipment and to the environment, requires no cleanup after use, is electrically non-conductive, and is very cost effective. The CO_2 is typically stowed under high pressure in 100-lb bottles in a locker or compartment remote from the protected space as shown in Fig. 28.7. In the event of a fire in a protected space, the system is manually activated. Vent fans serving the space are automatically shut down and dampers are automatically closed. Upon activation, a CO_2 driven alarm is sounded, giving anyone in the space a 20-second evacuation warning before the space is flooded with CO_2. Fig. 28.8 shows the alarm and warning. The CO_2, which is heavier than air, displaces the oxygen in the space and extinguishes the fire by smothering the combustion process. The danger that arises for people who fail to evacuate in time is that they too will be deprived of oxygen. Anyone in the space must recognize the CO_2 flooding alarm and know the locations of the exits. The CO_2 is discharged through fixed nozzles as shown in Fig. 28.9.

Another total-flooding fire-suppression medium, widely used in fixed systems until recently, was Halon. However, most Halon has been banned because of its ozone-depletion characteristics and its toxicity in fire conditions. Several replacement chemicals have been developed which also extinguish fire through a combination of cooling and chemical-based fire inhibition.

Aqueous film-forming foams (AFFF) are used for fighting oil-fueled fires, particularly on the open decks of tankers and in machinery spaces. Tankers of over 4,000 deadweight tons require foam monitors, as in Fig. 28.10, for projecting a stream of foam. The foam is mixed with seawater and supplied via the fire main. The foam is then used to blanket the area where oil is on fire or likely to collect and burn. The foam forms a sealing film to smother the fire and reduce the chance of reignition, or to prevent ignition of the oil. Foam systems may be fixed installations, with the concentrate stored in a tank and a pump to inject the concentrate into the fire main branch. For machinery spaces the mixture is then distributed to the bilges via fixed piping. For open decks of tankers the foam is distributed to the monitors and fire stations. In other applications the concentrate is stored in strategically located containers and a **pick-up tube** is fitted at the fire hose nozzle to draw the chemical into the hose stream, producing the foam blanket.

Portable fire extinguishers provide a rapid response to fires, enabling them to be fought when they are small and easier to contain. Portable extinguishers typically contain CO_2, but may also be filled with chemical foam, a soda-acid combination, or a dry chemical. Portable extinguishers are rated for use on different types of fires: **Type A extinguishers** are for ordinary combustible materials; **Type B extinguishers** are for flammable liquids.; **Type C extinguishers** are for electrical fires and contain a non-conducting agent. Portable dry-chemical extinguishers may contain a potassium-bicarbonate based dry chemical known as Purple K used to extinguish flammable and/or combustible liquid fires.

Ships carry **self-contained breathing apparatus** (SCBA) for emergency use. These air packs, shown in Fig. 28.11, can be used for any situation in which the immediate environment poses a threat to breathing. Examples of such threats include smoke, CO_2, and toxic gases. In Fig. 28.11, SCBAs are in use as personnel enter a closed compartment: even if no toxic gases are present, corrosion robs a space of oxygen, and crew can be overcome when entering. **Fire suits** and other fire-fighting gear are also carried on board most ships.

28.5 Miscellaneous safety equipment

Bilge flooding detectors in machinery spaces alarm at the engine room console and on the bridge to alert the crew to rising water levels. The flooding might be caused by breaching of the hull, but common sources include leaks in seawater piping systems, especially at hull piping penetrations and sea valves.

Emergency position indicating radio beacons (EPIRB) alert authorities to a ship's sinking. EPIRBs are buoyant electronic devices that are carried in float-free cradles to be released if a ship sinks. The EPIRB floats on the water surface and sends a radio signal with geographic coordinates to a satellite receiver so that authorities can initiate rescue efforts.

Gas detectors are fitted on special purpose ships. Tankers and chemical carriers are fitted with volatile organic compound (VOC) detectors. Roll on/roll off (Ro/Ro) vessels are fitted with carbon monoxide (CO) detectors to alert the crew of dangerous concentrations of vehicle exhaust gases.

Chemical suits are carried aboard tankers and chemical carriers to protect the crew in the event of cargo leaks or spillage.

Safety harnesses, as shown in Fig. 28.11, are used by crew entering void spaces with questionable breathing environments. Crew subject to extreme ship's motion accelerations or those climbing high ladders may also use harnesses.

Distress signals, typically parachute flares, are carried to alert nearby vessels or rescue personnel of the location of a ship in distress.

Line-throwing appliances, normally rocket propelled, are used to pass lines to other vessels to transfer safety or rescue equipment.

CHAPTER 29
SHIPBOARD PROCEDURES

29.1 Reporting aboard

A new crewmember reporting aboard will find that although he or she was expected, the time of arrival was tentative because of travel arrangements. Upon clearing gate security and climbing the ship's gangway, the new engineer should ask the gangway watch for directions to the Chief Engineer's office. In the absence of the Chief Engineer, the new engineer should seek out the First Assistant Engineer. The new arrival will be assigned to a watch or advised when to **turn to**. Sometimes night engineers take the night watches in port, however, the new engineer may be part of a large turnover of ship's complement and watches may commence immediately after joining the ship. A berthing space, and with it lifeboat and fire stations, will be assigned, and a brief description of the ship's current status and of the Chief Engineer's expectations will be given.

The new engineer will report to the Captain's office. There, paperwork will be completed, including forms for taxes and allotment forms for sending money home or into bank accounts. If the ship is sailing to foreign ports, the new arrival will **sign articles**, a contractual employment agreement.

Prior to departure, the new arrival must locate his or her assigned lifeboat and fire stations and become familiar with the associated responsibilities at each station.

29.2 Shipboard organization

Shipboard organization is illustrated in Fig. 29.1. At the top is the Captain or Master of the ship. The Captain has the ultimate authority on board, and has the ultimate responsibility for the ship, its cargo, and all personnel.

The ship's crew is organized into departments. On most merchant ships, the two main operational departments are the deck department and the engine department. Additionally, there are support departments such as the Steward's department and Purser's department, which can be sizable on passenger ships.

The **Chief Engineer** is in charge of the engine department and all of the ship's machinery. It is the Chief's duty to ensure that the correct number of licensed and unlicensed personnel are on board, that necessary arrangements are made for maintenance and repair, and that there is sufficient fuel and freshwater on board at all times. He or she ensures that plant equipment is safe and is operating properly and efficiently. The Chief delegates some of these duties, but retains the overall responsibility.

Directly under the Chief Engineer is the **First Assistant Engineer** (on US-flagged ships; in most other registries he or she is called the Second Engineer**)**, who is usually a day worker performing and supervising maintenance and repair duties. The First Assistant advises the Chief in ordering spare parts and consumables.

Where engine room watches are stood, the other **assistant engineers** are each in charge of a watch. Also on each watch will be one or two **unlicensed personnel,** depending on the degree of plant automation. Like the engineers, unlicensed personnel are certified by the USCG (or other regulatory body) for their competence. Where machinery is fully automated and watches are not routinely stood, the Assistant Engineers are day workers. The Assistant Engineers are responsible for the unlicensed personnel assigned to them, and must ensure that they perform their duties properly.

Other day-working personnel may be included in the engine department depending on the type of ship and the installed equipment. For instance a passenger ship may carry plumbers and electricians, while a refrigerated cargo vessel may carry refrigeration engineers. Older ships may carry watchstanding **oilers** and **firemen**, while newer ships would utilize **QMED's** (Qualified Members, Engine Department).

The lowest-ranking person in the engine department is the **wiper**, whose job includes general cleaning, painting, and other general-assistance duties, as assigned by the First Engineer

29.3 Watch standing and routine duties

Ships with continuous engine-room watches generally have three assistant engineers who share the watches. **Sea watches** are stood for four hours twice each day from 12-4, 4-8, and 8-12, with eight hours off between watches. Overtime, if required and authorized by the First Engineer, is normally worked between 0800 and 1700. In port, watches may be extended to eight hours on duty, with sixteen hours off, or per-diem engineers may be hired to give the ship's engineers more time off duty.

On a ship having a day-working First Engineer, the 12-4 and 8-12 watches are customarily stood by the Third Assistant Engineers and the 4-8 by the Second Engineer. This arrangement provides the Second Assistant with the longest daytime period for overtime work.

Prior to assuming duty for the first time, the new engineer must become familiar with the plant. It is also a good idea to relieve the watch early in order to ask the engineer on watch plant to go over some of the more important features of the plant. For example, the new engineer will need to know the pumps that are on-line, valve line-ups, governor adjustments, combustion-control adjustments, the fuel pressures at which burners are cut in or out, etc. As soon as time is found, the new engineer should trace out all pipelines and bypasses, and become familiar with equipment operation and emergency procedures. Location of fire-fighting equipment must be ascertained.

On fully automated ships, an assistant engineer is generally assigned as the duty engineer for 24-hours, on a rotating basis. The duty engineer has the responsibilities of a watchstanding engineer, but need not be continuously present in the engine room during his tour. The duty engineer will be required to make periodic rounds and will be the first to respond to alarms, which are directed to his or her quarters. The duty engineer is expected to rectify minor problems, but to call for assistance should the problem exceed his or her capabilities. While on-duty, the duty engineer is otherwise available for routine day work or leisure activities as long as these do not interfere with his/her responsibilities as the duty engineer.

Whether watches are stood or all engineers are on day work, the routine maintenance responsibilities are divided among the assistant engineers. One third assistant might be responsible for the lubricating-oil system including strainers and purifiers, while the other third assistant might be responsible for the distillers. Other auxiliaries requiring periodic maintenance might be divided between the third assistants. Traditionally, the second engineer is responsible for the boilers and fuel systems, while the first engineer is responsible for the propulsion machinery.

The duty of the watchstanding engineer is to maintain a safe and efficient plant. Watch routines include periodic rounds of the machinery spaces, usually alternating the rounds with the watch standing oiler or QMED. During these rounds the engineer visually inspects all equipment looking for abnormal conditions. The engineer will tighten leaking packing glands, verify that all pressures, temperatures, and fluid levels are in normal ranges, feel motors and bearings for temperature and vibration levels, and observe and listen for any abnormal condition.

The engineer on watch must be able to handle routine and emergency operations. He or she must be able to operate the propulsion machinery, and must maintain liquid levels, lubricating-oil and cooling-water flows, and keep pressures and temperatures within limits. He or she must be able to start, adjust, and stop auxiliaries. He or she must be well versed in emergency procedures, and be ready to initiate appropriate action before assistance arrives.

The watch engineer is expected to correct routine abnormalities without assistance, but if the problem is serious the Chief and/or the First Engineer are notified. If a problem begins to exceed his or her capabilities the watch engineer should ask for assistance from the Chief or the First. If a problem is escalating out of control, the **Engineer's Assistance Alarm** is actuated. A loud alarm will then sound throughout the accommodation, signaling *all* engineers to come immediately to the engine room to provide assistance. Major casualties require a team effort.

Depending on the extent of automation and automatic data recording, the unlicensed member of the watch may pump bilges, pressurize pneumatic tanks, record temperatures, pressures, and levels for entry into the log book, and performs other routine tasks. The engineer starts and secures the more complicated equipment that would not normally be started or adjusted by the QMED or oiler. A good working relationship with a competent QMED or oiler is important. The engineer is responsible for checking for instructions from the Chief or First, which may be verbal or written, as in the **Night Order Book**, and carrying them out. The engineer maintains the **Engine-Room Log Book**, recording important operating parameters. Recordings of performance parameters are used for trend analysis and to schedule maintenance to ensure continuous proper operation. The engineer also notes all non-routine operations and their time of occurrence in the log, which becomes a permanent and official record. At the end of the watch the engineer reports total revolutions and seawater temperature to the mate on watch for "dead reckoning" navigation and for the Bridge Log Book.

In **relieving a watch**, although the next watch officially starts on the hour, it is customary to relieve watches ten minutes before the hour as a matter of tradition, courtesy, and good seamanship. Arriving late for watch, even by one minute, is sufficient grounds for being reported in the log or discharged. The relieving engineer is expected to make a **pre-round** of the machinery spaces and steering gear before relieving the watch. With a good pre-round, the incoming engineer starts the watch aware of existing plant conditions. The outgoing engineer will give a detailed synopsis of the events that transpired during the previous two watches, report any equipment requiring special attention, note the current status of the plant and cite equipment that has been changed-over from operational to standby status, and point out any changes to the Chief Engineer's **standing orders of the watch.** Good seamanship and pride in the profession require that the outgoing watch turn over the plant in good condition, and for the outgoing engineer to assist the relieving engineer in correcting existing problems before leaving the engine room.

29.4 Firefighting and lifeboat procedures

Procedures for firefighting and the use of lifesaving systems are governed by regulatory bodies. Every member of the crew is assigned a **fire station** and a **lifeboat station**, and specific duties at each station, via a muster list posted in berthing spaces and public rooms. A new crewmember is required to be familiar with these assigned duties before his or her first voyage begins. As soon as possible but not later than two weeks after joining a vessel, a new crewmember is required to be given onboard training in the use of the equipment. Every crewmember must participate in one fire drill and in one boat drill every month.

29.5 Bunkering

Bunkering the ship is the refueling process required to fill the ship's storage tanks. Bunkering may be done once each voyage, or more frequently. The time between bunkerings is usually an economic decision made by the ship owner or manager on the basis of fuel cost, quality, and availability, and the ship's trade route, schedule, and storage capacity.

29.5.1 Bunkering system

Bunkering the ship makes use of the ship's fuel-oil transfer piping system to receive fuel under pressure from a barge or shore-side pumping station. Filling stations are usually located on each side of the ship, each fitted with double gate valves, approximately six inches (150 mm) in size, and a standard flanged connection, which is blanked-off when not in use. This valve/blank-flange arrangement is a safety measure to avoid accidental oil spills.

29.5.2 Bunkering operations

Procedures presented below may be followed for bunkering a typical ship. However variations occur. As an example, a cargo ship may have many smaller double bottom tanks and wing tanks, while a tanker may have fewer, but larger, deep tanks.

Preparation:

1. Formulate a filling plan ahead of time. For example, start filling the after port and starboard tanks and work forward.
2. Sound all tanks prior to bunkering. Prepare a table indicating each tank number, full tank capacity and sounding, starting tank levels and volumes, volume to be pumped, and start/end time.
3. Burn fuel from the settlers so that their level is low during bunkering. Generally, the piping is arranged with a standpipe that will overflow to the settlers if the line pressure exceeds the static head of the standpipe, to minimize the risk of an overflow to deck.
4. Install drip pans under all fuel oil tank vents in the event that the tank "burps" a slug of oil that is trapped in the vent piping as air is displaced during filling.
5. Install wooden plugs into each deck scupper. This precaution is taken to retain fuel that is accidentally spilled on deck, which can be cleaned up without polluting the waterway.
6. Inform the mate on watch that bunkering will be taking place, and at the approximate time. The mate will raise the "bravo" pennant, a red flag used as an international signal that the ship is bunkering.
7. Test and verify proper operation of the walkie-talkie or other ship-to-barge communication system.

Filling operation:

8. Line up the filling valves in the ship's transfer system. Ensure that all valves in the correct line up are opened and that all other valves are closed. Double check the valve line up before commencing taking bunkers.
9. It is considered good practice to attach a **static ground cable** or **bonding cable** between bare metal on the ship to bare metal on the barge, to avoid a static electric potential difference between the ship and barge that could result in a spark.
10. Rig and connect the filling hose from the filling connection on the barge to the ship's filling connection. Open the double gate valve in preparation for filling.
11. Begin filling slowly. The barge pumping rate is variable. Check all of the ship's fuel tanks as the operation begins to ensure that the fuel is entering the expected tanks and is not being pumped to other tanks. Sound and verify the tank levels and reconcile them against the barge meters. Check the deck vents for any evidence of oil on deck.

12. After ensuring that filling is going as expected, order the barge to increase the filling rate. The barge cost is based on a contracted bunkering time and any time beyond that will result in a higher fee to the ship's owner. However, spills must be avoided at all costs.
13. When the tanks are about 90% full, **topping-off** begins. Topping-off is the procedure where the tanks are filled to the final volume at a reduced rate. Tanks are often topped-off in port and starboard pairs. Top-off by opening the fill valves for the next set of tanks to be filled. As the level in the tank being topped-off rises, throttle its fill valve to further reduce the filling rate.
14. When the tanks being topped-off reach 95% full, close the filling valves to those tanks. Up to 5% **slack** is left in each tank to allow for expansion of the fuel when the oil is heated later for transfer. The near-full level is desired to limit the free-surface effects of fluid in tanks.
15. In the event that a settler high-level alarm sounds, or if any other abnormal condition is noted, immediately stop the filling operation. Stopping the filling operation is normally done by communicating with the barge to actuate their emergency trip of the pump. Only if their response is too slow and the abnormal condition is serious enough, should the ship's filling valves be closed at the filling station. Stopping the pump is preferred to closing the ship's valve to avoid a pressure rise on the barge pump, piping, and hose; although the barge system and the hose should be protected by relief valves, the actual condition of the equipment is not known.
16. During the filling process, take samples of the fuel. Record the names of the producer and vendor of the oil, its flash point and specific gravity. The flash point of the oil must be a minimum of 140°F (60°C). Estimate and record the heating value using a fuel-oil chart. Record the barge name or number, and the start and finish meter readings. The samples are retained on board until the fuel is consumed, or may be sent for tested ashore to protect the ship owner from receiving fuel of lower grade than contracted.
17. About ten or fifteen minutes before finishing, give the barge crew a standby signal. After the transfer is complete, the barge crew will blow the hoses clear of oil, after which they are disconnected.
18. After completing bunkering, sound all tanks and reconcile the resulting quantities with the barge soundings and fuel meter reading. When the oil is at a temperature other than 60°F (15°C) volume corrections may be made with the equation, $V_{60\,F} = V_{actual} + 0.0004(T\text{-}60°F)$.

29.6 Arriving the ship

When approaching a port, the mate on watch will call down to the engine room to inform the watch approximately an hour before arrival at the pilot station. The standing orders of the watch indicate who is to be notified. Generally the watch engineer calls the first assistant engineer who will take charge of maneuvering, while the watchstanders maintain the plant in good order.

Upon or before arrival, the evaporator will be changed over from filling the potable water tanks to fill the distilled water tanks to avoid contamination of the drinking water. The shaft revolution counter and time of arrival are noted in the engine room log book and the **bell book**. The plant is prepared for maneuvering, as described in earlier in chapters.

If the ship has a scoop injection for circulating water, the main circulator valves are opened and the main circulator started. The scoop injection is then closed. If the vessel is fully laden or entering a shallow water port, the sea suctions are shifted from low to high suctions in order to avoid picking up debris and silt from the harbor bottom. If the ship is fitted with bow or stern thrustors, a standby generator may be started to carry the extra load.

During maneuvering, for ships without bridge control of the main engine, commands for changes of shaft speed and direction, called **bells**, are received on the **engine-order telegraph**. Bells are answered promptly and logged into the engine room bell book along with the time of the bell, using the standard bell symbols shown in Fig. 29.2.

After **Finished With Engines** is rung down on the engine-order telegraph, equipment will be secured depending on the length and nature of the stay in port, and on plans to undertake maintenance and repair not feasible at sea. On steamships, a stay of more than a day may warrant securing one of the ship's boilers to reduce fuel consumption.

29.7 Preparing for shipyard work or drydocking

Periodically, a ship enters a shipyard for inspection or repair, or for routine hull cleaning and painting and major machinery maintenance. Because of the loss of revenue (the ship is said to be **off hire**) and the cost of berthing space, the Chief Engineer prepares detailed plans in advance to minimize yard time. These plans may include a list of required tasks to be accomplished, and decisions about work to be performed by the ship's complement and those to be performed by the shipyard or other specialists. The Chief Engineer tries to identify the **critical path** that will minimize the off-hire time. The ship's engineers, usually joined by a **port engineer** from the shipowner's office, will provide supervision and otherwise represent the shipowner's interests to ensure that all work is accomplished satisfactorily. If the ship is going into a **dry dock**, the Chief Engineer will furnish a **docking plan** to the shipyard, which indicates the dimensions, locations, and loads for the blocks that will support the ship in the dock.

During short shipyard visits the ship maintains its own power. Longer visits may require that the ship's plant be secured and that electric power and other utilities be obtained from shore connections. The ship therefore has connections for shore supply of electricity, potable-water, firewater, and steam, and for condensate return.

If the ship is to be **laid up** for an extended period with minimal utilities in climates where freezing may occur, machinery and piping must be heated or drained of water. If freezing is not a concern, the boilers can be laid up either wet or dry. If laid up wet, the boiler is filled with deaerated and chemically treated water until it issues from the air vent. If dry, the steam and water drums are opened for permit circulation of air, or a desiccant is placed inside the drums. To protect the firesides, a cap may be installed over the uptake outlet and a ventilating unit may be installed to blow heated air into the furnace. It is common practice to install infrared heat lamps on motors to keep them warm and the windings free from condensation. Gearboxes, turbine casings, engine crankcases, and control consoles may be fitted with fans and dehumidifiers. If the refrigeration system is to be secured, the refrigerant is pumped down to the receiver, which is then isolated by closing the receiver inlet and outlet valves tightly.

29.8 Departing the ship

Before departing, secured equipment is started and the plant is otherwise made ready for sea. The mate on watch is called before turning the propeller shaft to confirm that there are no mooring lines in the water that may foul the propeller. For the same reason, when the mates are ready to "let loose the lines," they call down to the engine room and the engines are stopped. After the lines are retrieved, the ship maneuvers out of the port. On ships without bridge control of the main engine, the engineers answer the bells rung down on the engine-order telegraph. While maneuvering, the First Assistant or Chief Engineer may take charge of the watch, while the watch engineer makes adjustments to compensate for constantly changing conditions. Each bell and its time are recorded in the bell book.

When the ship clears the pilot station, the mates call down indicating departure of the ship. At departure, the ship is slowly brought to sea speed while all plant conditions are closely monitored. As speed rises, the plant is shifted from maneuvering mode to sea-going mode as described in previous chapters. The departure time and shaft revolution counter are recorded in the bell book.

29.9 Oil record book

In the **Code of Federal Regulations (CFR)** 33 CFR Chapter I Subchapter O - Pollution contains the United States laws that govern the discharge of oil-containing water into navigable waters. Similar regulations apply to ships of other registries. These regulations limit the amount of oil permitted in the bilge and ballast discharge effluent to an oil content of less 15 ppm (parts of oil per million parts of effluent) if the vessel is within 12 nautical miles of the nearest land or a limit of 100 ppm if the vessel is outside these limits. It is also required that the vessel maintains an oily-water separator. The CFR lists exceptions for emergencies, requirements for reporting oil observed over the ship's side, and the fines and other penalties for failing to report an occurrence.

To control and minimize pollution, the regulations require that the vessel maintain an **Oil Record Book**. The Oil Record Book is property of the regulating authority. It is required for all tankers of 150 gross tons or more, all ships of 400 gross tons or more, and manned oil rigs. Separate Oil Record Books are maintained on tankers and on other vessels carrying more than 200 cubic meters of cargo oil for the machinery spaces and for the cargo/ballast operations.

Among the entries in the Oil Record Book are:

Ballasting or cleaning of fuel oil tanks:

1. Identity of tank(s) ballasted.
2. Whether cleaned since they last contained oil. If not, the type of oil previously carried.
3. Position of the ship at the start of cleaning.
4. Position of the ship at the start of ballasting.

Discharge of dirty ballast or cleaning water used in cleaning fuel oil tanks:

1. Identity of tank(s).
2. Position of the ship at the start of discharge.
3. Position of the ship at the completion of discharge.
4. Ship's speed during discharge.
5. Method of discharge
    a. 100 ppm equipment,
    b. 15 ppm equipment,
    c. or the name of the reception facility and the port of discharge.
6. Quantity discharged.

Disposal of oil residues and sludges:

1. Quantity of residue retained on board for disposal.
2. Method of disposal
    a. the name of the reception facility and the port of discharge,
    b. mixed with bunkers,
    c. transferred to another tank, with tank identities,
    d. any other method, which must be stated.

Non-automatic discharge overboard of bilge water that has accumulated in the machinery spaces:

1. Quantity discharged.
2. Time of discharge
3. Position of ship at start and completion of discharge.
4. Method of discharge

a. 100 ppm equipment,
b. 15 ppm equipment,
c. or the name of the reception facility and the port of discharge,
d. to identified slop or collecting tanks.

Automatic discharge overboard of bilge water that has accumulated in the machinery spaces:

1. Time when the system is placed into the automatic mode for overboard discharge.
2. Time when the system is placed into the automatic mode for an identified collecting of slop tank.
3. Time when the system is placed in manual mode.

Each entry in the Oil Record Book must be signed by the person(s) in charge of the operation.

INDEX